Asia's Alliance Triangle

ASAN-PALGRAVE MACMILLAN SERIES

The Asan Institute for Policy Studies is an independent, non-partisan think tank with the mandate to undertake policy-relevant research to foster domestic, regional, and international environments that are conducive to peace and stability on the Korean Peninsula and Korean reunification.

The Asan Forum

In June 2013, the Asan Institute for Policy Studies launched *The Asan Forum*, an online journal dedicated to the debate and analysis of issues that affect Asia and beyond. Under the leadership of Gilbert Rozman, *The Asan Forum* brings together scholars and policy experts from across the region and disciplines, seeking a diversity of views to harness the intellectual synergy created when perspectives compete and, more importantly, complement one another.

OTHER BOOKS IN THE SERIES

Gilbert Rozman (ed.), *China's Foreign Policy: Who Makes it and How is it Made?*, 2013

Clement Henry & Jang Ji-Hyang (eds.), *The Arab Spring: Will it Lead to Democratic Transition?*, 2013

Bong Youngshik & T.J. Pempel (eds.), *Japan in Crisis: What will it Take for Japan to Rise Again?*, 2013

Mo Jongryn (ed.), *Middle Powers and G20 Governance*, 2013

Mo Jongryn (ed.), *MIKTA, Middle Powers, and New Dynamics of Global Governance*, 2014

Baek Buhm-Suk & Ruti G. Teitel (eds.), *Transitional Justice in Unified Korea*, 2015

Gilbert Rozman (ed.), *Asia's Alliance Triangle: US-Japan-South Korea Relations at a Tumultuous Time*, 2015

Asia's Alliance Triangle

US-Japan-South Korea Relations at a Tumultuous Time

Edited by
Gilbert Rozman

palgrave
macmillan

First published in 2015 by
PALGRAVE MACMILLAN®

The authors have asserted their rights to be identified as the authors of this work in accordance with the Copyright, Designs and Patents Act 1988.

Palgrave Macmillan in the UK is an imprint of Macmillan Publishers Limited, registered in England, company number 785998, of Houndmills, Basingstoke, Hampshire, RG21 6XS.

Palgrave Macmillan in the US is a division of Nature America, Inc., One New York Plaza, Suite 4500, New York, NY 10004-1562.

Palgrave Macmillan is the global academic imprint of the above companies and has companies and representatives throughout the world.

Hardback ISBN: 978–1–137–54170–3
E-PUB ISBN: 978–1–137–54172–7
E-PDF ISBN: 978–1–137–54171–0
DOI: 10.1057/9781137541710

Distribution in the UK, Europe and the rest of the world is by Palgrave Macmillan®, a division of Macmillan Publishers Limited, registered in England, company number 785998, of Houndmills, Basingstoke, Hampshire RG21 6XS.

Library of Congress Cataloging-in-Publication Data

Asia's alliance triangle : US-Japan-South Korea relations at a tumultuous time / edited by Gilbert Rozman.
    pages cm.—(Asan-Palgrave Macmillan series)
    Includes bibliographical references and index.
    ISBN 978–1–137–54170–3 (hardback)
    1. United States—Foreign relations—Japan. 2. Japan—Foreign relations—United States. 3. United States—Foreign relations—Korea (South) 4. Korea (South)—Foreign relations—United States. 5. Japan—Foreign relations—Korea (South) 6. Korea (South)—Foreign relations—Japan. I. Rozman, Gilbert.
E183.8.J3A864 2015
327.73052—dc23   2015013625

A catalogue record of the book is available from the British Library.

Contents

Illustrations

Figure

Tables

Acknowledgments

This is the first book drawn from the postings in *The Asan Forum*, an online journal started by the Asan Institute of Policy Studies in June 2013. Focused on international relations in the Asia-Pacific region, this journal provides continuous coverage of the important bilateral and multilateral relationships reshaping East Asia. In this series, we group selected postings not by when they appeared but into distinct arenas that have each been rapidly transforming in the 2010s, as seen from diverse angles in the region and in the West. The selections in this book come from the first nine issues of this bimonthly journal, centering on the US-Japan-South Korea alliance triangle.

Many persons have made this book possible. Hahm Chaibong, president of the Asan Institute, established the journal and set its overall direction in support of academic standards and balanced regional coverage. The staff of the journal has devoted its energy to ensuring timely, sustained posting of high-quality materials. The authors who write for the journal have been very cooperative in taking this task seriously and meeting deadlines. We are all appreciative of the support Farideh Koohi-Kamali of Palgrave Macmillan has given to this endeavor to turn online postings into books.

Gilbert Rozman
Editor in chief of the Asan Forum
December 2014

Introduction

Gilbert Rozman

A key to understanding the transformation under way in East Asia is to reconsider what we thought we knew about the US-led alliance triangle with Japan and South Korea. The prevailing assumption has been that these three countries are drawing closer due to shared strategic interests, values, and attitudes toward economic and international cooperation. It was long expected that the close US alliances with the two would be transformed into a trilateral alliance—the backbone of a network of US alliances and defense partnerships along the Pacific Ocean and extending to the Indian Ocean. With North Korea's threat potential rising precipitously and growing uncertainty over China's strategic intentions and, recently, Russia's "turn to Asia," conditions seemed ripe for Japan and South Korea to narrow their differences. The period 2013–2014 covered in this book exposes the fallacy in these arguments and is worthy of close scrutiny for what it tells us about prospects for an alliance triangle.

No topic drew greater attention in the first year of the *Asan Forum* than US-ROK relations. A close second was ROK-Japan relations. Also important was the changing approach in Japan to national security with strong implications for its relations with the United States. A special feature of the journal has been its attention to national identities and their impact on international relations, including identities in Japan and South Korea. Pulling coverage of these themes together in one book, by picking and choosing from materials in the first nine issues of the journal, results in the most up-to-date, informative presentation available on the US alliance triangle.

In 2013–2014 the US-Japan-ROK triangle experienced the most uncertainty in its more than 60 years as a force in the security architecture of East Asia. It was troubled by poor Japan-ROK relations, clashing Japanese and ROK approaches to China, and divergent assessments of challenges from North Korea and Russia. US relations with Japan and South Korea appeared to remain strong, but new undercurrents did not prove to be readily manageable. This short time span warrants intense attention with diverse perspectives from all three sides of the triangle. This book responds with a continuous record of the period, prepared by specialists who are among the most knowledgeable, delving deeply into the timeliest and most significant issues, and conveying conflicting opinions on how best to address some critical problems.

This is a running commentary of observations and analyses from the middle of 2013 to the end of 2014. They show how challenges were being faced as they kept arising. By the start of 2015—the seventieth anniversary of the end of World War II—the overall situation had become quite clear. The Obama administration would build on the momentum of its 2014 efforts to bridge the gap between the Abe and Park administrations, at least to convince each side to avoid any moves that would exacerbate their problems over historical memory. Planning for summits in the United States would focus on preventing them from being treated as a zero-sum game; the US objective is two stronger alliances and increased trilateralism. Given the growing concern in both Tokyo and Seoul that their quarrel is damaging their strategic interests, one could expect limited success from US diplomacy. Yet, even a somewhat upbeat mood ahead should not turn our eyes away from two tortuous years in Japan-South Korea relations and US frustrations in response. The record of this most troublesome gap between the urgency of alliance triangulation and the reality of bilateral discord and distrust in one leg of the triangle will be of enduring salience.

The Five Parts of the Volume

The Asan Institute convened three conferences on the state of US-ROK relations at various times from June 2013 to the fall of 2014. Synopses of these and a separate article by Choi Kang, a leading authority in Korea, comprise part I of the manuscript. The synopses are not dry renditions of who said what or how each panel proceeded, but provide thematic interpretations of what of lasting relevance was said as part of exchanges of opinion. They capture the high quality of the discussions, drawing out differences of opinion and far-reaching conclusions that even participants at these gatherings may not have noticed without attending many panels and reviewing the notes from each. The result is a rich, bilateral perspective on a critical US alliance.

Roughly equal space is devoted in part II to the Japan-ROK relationship, approached in a different manner. Its components are two articles and a series of three Topics of the Month articles by another author. This produces a running commentary of what was interfering with the relationship and what proposals arose to improve it. One theme is the parallel effort during the first part of 2014 by Japanese and Korean officials and academics in Washington to persuade Americans of their point of view with skepticism raised in the audience toward presenters from both countries. Competition between Japan, the "cornerstone," and South Korea, "the lynchpin," over which is the "model ally" and whose agenda takes priority is likely to continue.

Part III focuses on Japan's national security policy, since that produced the liveliest debates of any internal matter affecting triangular relations. It includes articles by three Japanese experts, whose views are not fully in accord and reflect shifts over the full year and a half covered in the manuscript. It concludes with different views of how close Japan-US relations were in late 2014, covering both security and values.

Part IV starts with Obama's meetings with Park and Abe in the spring of 2014, a critical time when the United States was striving to repair the worst rupture in ROK-Japan relations. There is also some follow-up, covering later months. Authors from all three countries are represented. This part of the book also includes later commentaries— from the United States, Japan, and South Korea—addressing whether Japan and the United States share the same values and overall view of history. This was a persistent concern through 2014.

Part V adds the perspective of national identities in Japan and South Korea. Given how much of the discussion of troubled relations has centered on historical memory and other identity themes, it is fitting that we probe deeply into this dimension. The chapters explore public opinion, newspaper debates reflecting clashing schools of thought, and views on different bilateral relations and international relations issues.

Readers interested in international relations in Northeast Asia will find nothing like this book. It strikes a balance in covering viewpoints in three countries. It focuses on timely, recent developments at a critical juncture for the region. This is not a mix of unconnected articles, but pieces commissioned or written according to an overall plan or a series of interconnected plans. The whole is greater than the sum of the parts since the US alliance system serves as the overall theme. Linkages between national security, national identity, and regional diplomacy are made throughout.

These subject areas related to the alliance triangle fit together well as we look back on a tumultuous time, when historical memories troubled Japan-ROK relations, just as North Korea was becoming more threatening and China more aggressive. In Seoul, management of the North Korean danger was primary, while in Tokyo much greater attention focused on management of the China threat. Washington faced a challenge in steering trilateral relations forward while overcoming clashing national identities and more divisive interpretations of national interests. Each part of this book sheds light on one or another side of the struggle to solidify alliances, boost security triangularity, counter unprecedented challenges, and face external complications.

Bilateral relations among the three countries could not escape the shadow of their essential triangularity. Whether it was collective self-defense by Japan (welcomed by the United States but troubling to South Koreans), historical memory centered on "comfort women" in South Korea (accepted by the United States but at odds with the growing emphasis on historical pride in Japan), or prioritizing containment of North Korea or China (increasingly divisive to Seoul and Tokyo), this book captures a time of unprecedented testing of the alliance triangle. Reading these pieces in one batch serves a wider goal than following the online postings monthly. It offers an overview of triangularity, caught between growing urgency and deepening fissures. Only from this multiangled, sustained examination can we appreciate the rising salience of this triangle safeguarding security in Northeast Asia in a decisive transformation.

On the one hand, the US alliances with Japan and South Korea are strengthening, as overwhelmingly desired by the public in both countries and their leaders. This book captures the views of security specialists who prize these bilateral alliances in times of increasing regional instability. Planning for new contingencies is advancing fast. Expanded types of military cooperation are readily observed. Also, the geographical scope of each alliance is being broadened. As many articles indicate, 2013–2014 may rightly be called the heyday of the two principal US alliances in Asia. On the other hand, there not only have been troubling questions about triangularity, focusing on the Japan-ROK leg, but also deep-seated uncertainties over management of China and its implications for an obsessive objective in both Japan and South Korea. The obsession for the administration of Abe is historical revisionism, arousing openly expressed disappointment in the United States with how this is interfering with a realist foreign policy in the face of security threats. For the Park Geun-hye administration, priority centers on North Korea—both as a threat and as brethren of the same nation—and on China's unparalleled influence on both diplomacy and pressure to deal with that country. Problems deepened over both these obsessions and alliance triangularity.

This book offers snapshots of the debates and perceptions within Japan and South Korea during a tumultuous year. The Japanese and Korean authors reflect thinking as it was evolving in response to rapidly changing developments. *The Asan Forum* is committed to presenting perspectives on international relations within East Asia. In this case, the often contrasting viewpoints of Japanese and South Korea specialists as well as officials and media make a compelling case for juxtaposing their thinking. In this way, we delve deeply into a vital triangle, which many see through the prism of US policy and analysis without appreciating the challenges coming from its allies.

The Asan Forum includes country reports (with bimonthly coverage of writings in South Korea and Japan, among other countries), regular reports on seminars in a column titled "Washington Insights," and articles grouped in sections such as Special Forum, Open Forum, Topics of the Month, and National Commentaries. Sometimes articles incorporated into this book refer to other pieces in the journal. Readers may want to consult the online journal to find these references or to look for additional information on the themes covered herein. The chapters in this book have, in some cases, been slightly modified to avoid unnecessary cross-references to the journal and for consistency.

PART I

The US-ROK Alliance at 60 Years

CHAPTER 1

Synopsis of the Asan Washington Forum, 2013: "The Enduring Alliance: Celebrating the 60th Anniversary of ROK-US Relations," Parts 1 and 2

Gilbert Rozman

The Asan Washington Forum, June 24–25, 2013

Agenda

Day 1

Table 1.1   The Asan Washington Forum, June 24–25, 2013: Day 1

Day 1: Monday, June 24, 2013

Time	Place	Panel title	Panelist	Affiliation
08:00–09:00	*Amphitheater foyer*	Registration		
09:00–10:30	Opening ceremony *Amphitheater*	Welcoming remarks	Hahm Chaibong	The Asan Institute for Policy Studies
		Opening remarks	Chung Mong Joon	The Asan Institute for Policy Studies
		Keynote speech	Richard Cheney	Former vice president of the United States
10:30–11:00	*Amphitheater foyer*	Break		

continued

Table 1.1 Continued

Day 1: Monday, June 24, 2013

Time	Place	Panel title	Panelist	Affiliation
Session I 11:00–12:00	*Amphitheater*	Sixty years of the alliance	(Mod) Hahm Chaibong	The Asan Institute for Policy Studies
			Burwell B. Bell	Former commander, US Forces Korea
			William Cohen	Former senator and secretary of Defense
			Han Sung-Joo	Former minister of foreign affairs, ROK
			Park Jin	Former member, National Assembly, ROK
12:00–14:20	*Atrium ballroom*	Lunch dedicated to Korean War Veterans Speech by John Warner (Former US senator, R-VA)		
Session II 14:20–15:20	*Amphitheater*	The Alliance and North Korea	(Mod) David Sanger	*The New York Times*
			Kim Sung-han	Former vice minister of foreign affairs, ROK
			Michael O'Hanlon	The Brookings Institution
			Gary Samore	Belfer Center for Science and International Affairs
			Walter Sharp	Former commander, US Forces Korea
			Yu Myung Hwan	Former minister of foreign affairs, ROK
15:20–15:40	*Amphitheater foyer*	Break		
Session III 15:40–16:40	*Amphitheater*	The future of the alliance	(Mod) Choi Kang	The Asan Institute for Policy Studies
			Kil Jeong Woo	National Assembly, ROK
			Lee Chung Min	Yonsei University
			Mark Minton	The Korea Society
			Douglas Paal	Carnegie Endowment for International Peace
			Paul Wolfowitz	Former US deputy secretary of defense

Table 1.1 Continued

Day 1: Monday, June 24, 2013

Time	Place	Panel title	Panelist	Affiliation
16:40–17:00	*Amphitheater foyer*	Break		
Session IV 17:00–18:00	*Amphitheater*	The alliance and the future of East Asia	(Mod) David Rennie	*The Economist*
			Graham Allison	Belfer Center for Science and International Affairs
			Richard Bush	The Brookings Institution
			Kurt Campbell	The Asia Group, LLC
			Hahm Chaibong	The Asan Institute for Policy Studies
			Joe Lieberman	Former US senator (I-CT)
18:00–18:40	*Atrium ballroom*	Reception		
18:40–21:00	*Atrium*	Gala dinner		
		Dinner speech by Madeleine K. Albright, former secretary of state		

Day 2

Table 1.2 The Asan Washington Forum, June 24–25, 2013: Day 2

Day 2: Tuesday, June 25, 2013

Time	Panel title	Panelist	Affiliation
08:30–09:00	**Registration**		
Session I 09:00–10:20	The Ambassadors' Dialogue: challenges for the alliance	(Mod) Hahm Chaibong	The Asan Institute for Policy Studies
		Han Sung-Joo	Former minister of foreign affairs, ROK
		Christopher Hill	University of Denver
		Thomas C. Hubbard	McLarty Associates

continued

Table 1.2 Continued

Day 2: Tuesday, June 25, 2013

Time	Panel title	Panelist	Affiliation
10:20–10:30	**Break**		
Session II 10:30–11:40	Public opinion: alliance, security, nukes	(Mod) J. James Kim	The Asan Institute for Policy Studies
		Charlie Cook	The Cook Political Report
		Kim Jiyoon	The Asan Institute for Policy Studies
		Bruce Klingner	The Heritage Foundation
		William Tobey	Belfer Center for Science and International Affairs
11:40–11:50	**Break**		
Session III 11:50–13:00	Dealing with a nuclear North Korea	(Mod) Shin Chang-Hoon	The Asan Institute for Policy Studies
		Lee Chung Min	Yonsei University
		Michael O'Hanlon	The Brookings Institution
		Bennett Ramberg	Foreign Policy Consultant and Writer
		Yamaguchi Noboru	National Defense Academy of Japan
13:00–13:50	**Lunch**		
Session IV 13:50–15:00	Dealing with North Korea's human rights	(Mod) Baek Buhm-Suk	The Asan Institute for Policy Studies
		Roberta Cohen	The Brookings Institution
		Frank Jannuzi	Amnesty International USA
		Kil Jeong Woo	National Assembly, ROK
		Marcus Noland	Peterson Institute for International Economics
15:00–15:20	**Break**		
Session V 15:20–16:30	The virtual alliance	(Mod) Walter Lohman	The Heritage Foundation
		Michael Auslin	American Enterprise Institute
		Bong Youngshik	The Asan Institute for Policy Studies
		Nishino Junya	Keio University

Table 1.2 Continued

Day 2: Tuesday, June 25, 2013

Time	Panel title	Panelist	Affiliation
16:30–16:50	**Break**		
Session VI 16:50–18:00	Korea between US and China	(Mod) Hahm Chaibong	The Asan Institute for Policy Studies
		Choi Kang	The Asan Institute for Policy Studies
		Thomas Christensen	Princeton University
		Bonnie Glaser	Center for Strategic and International Studies
		Gilbert Rozman	Princeton University
		Zhao Quansheng	American University
18:00–18:10	**Closing remarks**		

The Asan Institute recently held its inaugural forum in the United States, "The Enduring Alliance: Celebrating the 60th Anniversary of ROK-US Relations." This synopsis provides an opportunity to assess the conundrum of a bilateral relationship that has never been in better shape than over the past several years, which also faces some of the most serious uncertainties in six decades about its capacity to manage new challenges. A glance at the conference schedule will indicate who participated on which panels. The names should be familiar for their expertise to those who are following bilateral relations between the United States and South Korea. In this synopsis who said what is not of concern nor is the order of the discussion. Instead, the objective is to transmit the essence of the arguments on themes recurrent in the discussion. We are not interested in conveying a consensus, but in stimulating deeper understanding and discussion based on informed arguments.

The Asan Washington Forum was an exercise in scrutinizing an alliance from diverse angles, geographical, chronological, and thematic. The US-ROK dyad was viewed from the perspective of multiple triangles, especially those involving North Korea, China, and Japan. This synopsis is the first of two presentations of the arguments drawn from the forum. Much of the discussion of North Korea in relation to the US-ROK alliance is reserved for the second part. North Korea was understandably a major consideration throughout the conference; it also appears often in this synopsis. Yet, when triangles become the focus, the emphasis below is primarily on China and secondarily on Japan, not on North Korea.

One characterization of the purposes of the forum was "Remembrance, Reflection, and Renewal." Although there were moving memories of the Korean War shared with the audience and informative recollections of ups and downs in the evolution of the alliance brought to the fore by military leaders and diplomats, it is mainly the lessons from past experiences that draw our attention. Reflections from many angles on the state of current relations serve to identify some principal challenges that are being faced at the present crossroads for the alliance and for the region. The notion of renewal rests not only in regional attention to the zodiac cycle of 60 years, but also on clear-eyed awareness that a different environment has emerged, which requires revitalization of the alliance agenda. Much discussion centered on what has changed and what responses best meet new challenges.

This was understandably not an occasion for doubters on the value of the alliance to take the stage. Rather there was talk of great sacrifices made during the Korean War for a just cause, gratitude by Koreans for the sustained US commitment, pride in the remarkable achievements by Koreans, and satisfaction felt by Americans for one of their country's crowning successes in international relations. The image of the Republic of Korea conveyed is of a model of global significance today, economically, politically, and culturally. This sense of contentment with what some recognized as a model alliance did not stand in the way of forthright, often critical, scrutiny of problems, past and present.

In judging the alliance, many viewed its enduring strength and current closeness as proof of remarkable success. After all, it deterred a North Korean war and recurrent attempts to destabilize South Korea. A dissenting assessment was that North Korea's developing nuclear threat is evidence of failure. Other criteria were introduced, such as the success in 2009–2012 in broadening the scope of the US-ROK partnership in regional and global affairs to an exceptional degree or the frustrating failure to realize the US objective of a triangular alliance with Japan, causing more serious problems in the current, troubled era.

The first day of the conference was bookended by contrasting keynote addresses. In the remarks by former vice president Dick Cheney and former secretary of state Madeleine Albright, participants heard two strikingly divergent assessments of how the alliance has dealt with North Korea over the past two decades with clear lessons for the period ahead. Cheney attacked wasting time trying to persuade North Korea, while he made a strong case for unilateralism and reassurance to allies through strength. Albright argued instead for multilateralism or partnerships, coordinating with South Korea and patiently staying the course, as she thinks would have been advisable after her visit to Pyongyang in 2000 when negotiations held some promise. The alliance continues to be tested by responses to North Korea, but increasingly it is also evaluated as part of a broader, regional context.

While the history of the alliance appeared primarily through recollections of the sacrifices made in the Korean War and occasional references to troubling challenges such as Jimmy Carter's plan to pull US troops out of South Korea and the difficult relationship of Roh Moohyun and George W. Bush, history was never far below the surface of the forum discussion. With the goal of drawing lessons in facing new complexities, former officials pointed to how difficulties in the 1990s and 2000s were addressed and then overcome.

The alliance was evaluated in terms of its diplomatic and military dimensions as well as additional interest in its economic and cultural ones. Over the course of two days of panels, the alliance's prospects in a rapidly changing security environment stayed clearly in the forefront. One concern in discussions over how to strengthen the alliance was what changes in military ties are desirable and for what purpose. Whether the issue was burden sharing, missile defense, or the reintroduction of nuclear weapons into South Korea, there were differences over the impact: on China, on nonproliferation goals, and on how to apportion the cost to the two allies. However, diplomacy took center stage in most panels.

The role of values in the US-South Korea alliance was a recurrent theme. The US defense of freedom, South Korea's success in becoming the most vibrant democracy in Asia, the increasingly shared emphasis on support for universal values, and the growing advocacy for human rights in North Korea were all highlighted. When North Korea has closed the door to denuclearization, as in 2009–2013, stress on human rights proves easier, especially with conservatives in power in Seoul. A panel on human rights showcased the consensus achieved, bolstered by developments in the United Nations Human Rights Council in 2013. Talk turned to ways to empower the North Korean people or convince the regime that economic assistance is only possible after it changes its policies in this area. Yet, there was awareness that the South Korean public is divided about this priority, China is opposed, and the US priority is denuclearization. Putting stress on human rights may appear to be consistent with the cause of denuclearization at a time when diplomatic options are not working, but a real test awaits a new stage in diplomacy or in the North's isolation. On the one hand, discussions of North Korean "crimes against humanity" and other human rights violations drew sympathetic understanding on all sides. On the other, it aroused frustrations from those who saw others not giving this issue full priority and from some who doubt that idealistic aspirations will really outweigh strategic priorities.

The Triangle with China

Considerable analysis centered on the Sino-US-ROK triangle. One perspective held that when Sino-US relations improve the door opens for South Korea to play a more active role with China. The fact that an agreement in principle on making denuclearization the priority in dealing with North Korea seemed to be reached at the Obama-Xi Sunnylands summit made it easier a few weeks later for Park Geun-hye to pursue China on North Korea. Indeed, with Obama appearing to give the green light to Park to take the initiative, the argument was made that successful Sino-South Korean talks will serve US interests. Rather than a zero-sum triangle, it follows, the alliance plus China is a possible core of a regional security system with a trusted US ally gaining a pivotal role. This could even carry over after reunification that accepts South Korea as the dominant force in uniting the peninsula, some optimists argued. Yet, other views of China's calculus disagreed.

No matter what the subject of a panel might be, discussion kept winding its way back to China, whether as a passive responder that needed to be prodded into a greater role or as the real driving force in the region. Suggestions varied on how to deal with China: trust, persuasion, avoidance, or pressure. A China complex had taken root in Seoul, mixing fear of being left alone with China should the US role be reduced by deeper involvement in the Middle East and recurrent hopes that if no offense is given to Beijing there may be a way to anticipate its reactions and utilize them to Seoul's advantage. Hopes were boosted by signs of Chinese dissatisfaction with Pyongyang. It was suggested that North Korea over much of the past decade has been the number one subject of Sino-US talks, driving ups and downs in bilateral relations and buffeting the alliance as well. For Washington, as for Seoul, Beijing's handling of Pyongyang became the primary test of China's rise. China gained favor by starting the Six-Party Talks in 2003, helping to realize their one real achievement through the disablement of the North's reactor in 2007–2008, and showing some seriousness about the North's aggression in late 2010. Yet, when China put the blame on Lee Myung-bak for problems in 2009–2010 and was slow to respond to Kim Jong-un's bellicosity in late 2011 and early 2012, the opposite effect occurred. Those favoring trust in China or persuasion calculated that Beijing has now reconsidered this stance.

Another point of view is that China is more attentive to pressure than persuasion, as in signs of military preparedness to resist North Korea or counter its arms build-up. This is what is perceived to have occurred in early 2003, late 2010, and the spring of 2013. The key to China's accommodating responses on those occasions, some argued, was US-South Korean preparations for a military response, as in exercises or shows of force.

Whereas some participants saw promise in China's recent tougher posture toward North Korea, others saw a more nuanced Chinese response. One commentator noted that even as China has squeezed the North economically in some limited ways, it has also directed considerable funds to groups it favors. In this way, China seeks to shape the outcome of the internal transition as power is consolidated, pressuring the North to accept its plans for negotiations as well as to forge a leadership that would follow China's agenda.

Some were rather pessimistic, suggesting a tug-of-war in which first China and then the United States would beseech the South Korean government to adopt a posture in keeping with its strategy. On military matters, North Korea's growing nuclear force would lead to a strengthened US counternuclear strategy, making deterrence more credible, perhaps reintroducing tactical nuclear weapons into South Korea. Pessimists point to evidence that Beijing remains opposed to multilateralism on security in Northeast Asia that bypasses Pyongyang. It has repeatedly rejected 6–1 or five states meeting without North Korea and is keen on weakening the US-ROK alliance rather than acting in a manner that might enhance it. In this view, Seoul's pursuit of Beijing is likely to come at the expense of the alliance. Only by solidifying US ties and showing China that North Korea exacts a price on China's security is Seoul likely to have a desired impact. This seemed to be understood in 2010–2012, even by progressives, but the temptation has clearly been growing to rely on China again after Beijing hardened its tone toward Pyongyang.

Another proposal for breaking out of the 20-year cycle in talks with North Korea was for comprehensive reengagement. It called for overcoming the North's belief that nuclear weapons are essential for regime survival by convincing it that survival is not in doubt, while also making sure that there is no wedge between Washington, Seoul, and Tokyo as consultations with Beijing keep it as close as possible. Critical to this approach is an intense effort to convince China that a nuclear North Korea is not in its interest. Hopeful that such a recalculation is already under way, to the point that China is coming to regard a nuclear North as just as bad as collapse, optimists argued that this is occurring because China has realized that it had underestimated the serious negative impact of the North's nuclear obsession, including on closer alliance ties between Japan and the United States. On this basis, the recommendation holds, a peace regime can be pursued only if the Six-Party Talks show progress. Essential for the alliance under this sequence is to convince the South Korean public of the enduring value of the alliance beyond deterrence after reunification is realized. This line of thought sees China winning trust by its changing approach to North Korea and, in the process, South Koreans deemphasizing the alliance.

One problem with the various diplomatic strategies is that they rely on persuading China as if its leaders are not guided by a strategy that leaves little room for change of this sort. There is a tendency to overrate diplomacy when consensus is missing. In some of these strategies China is to be persuaded by increased flexibility from Seoul with US support. In others it is to be confronted with a more resolute allied stance to pressure the North, convincing China that there is only one path forward. Missing is compelling evidence that either of these approaches has much chance, given the way Chinese officials and analysts assess the situation and how Sino-North Korean relations will likely evolve.

Much depends on China's strategy for resuming the Six-Party Talks, persuading South Korea first that this is desirable. It would mean putting denuclearization in a broad and multistage context, conducive to China's objectives of downplaying the role of values, setting a course for regional security that limits the alliance role as in missile defense and conditioning reunification, if at all, on a North-South balance. Although many discussed the triangle without delving into China's strategic thinking, some warned of this outcome.

On the assumption that the North Korean threat is much diminished or that there has been reunification, some attention was turned to what should be done in order to reinvigorate the alliance as a force for regional stability. Many may doubt the need for it. China may press for its dissolution. Memories of persuading Mikhail Gorbachev of the wisdom of avoiding a neutral Germany stir hope that similar persuasion may work on China's leadership. The suggestion was made that some think tanks make the case for regional security based on the alliance and explain it to Chinese experts, as if that stands a chance of success. There is no reason to expect that US leverage over China is similar to what it had in dealing with Gorbachev a quarter century earlier, one discussant responded.

It is striking how often discussions of the alliance turned into triangular coverage of China. Occurring on the eve of Park Geun-hye's visit to Beijing, the forum foresaw maneuvering ahead that could impact the alliance. A common refrain was that if there is distance between the positions of the two allies, China wins, but that fell short of a policy recommendation. Concern that Obama was moving too quickly at his summit with Xi earlier in June led to advice that a quick China fix might damage the long-term US alliances. Another viewpoint held that South Korea embodies the best in how freedom and prosperity are built on the foundation of the US alliance system; so it reciprocates the US commitment to it with support for the liberal international order and is bound to do so in its appeals to China to be a responsible international actor, as in joining against actions of North Korea, whose crimes against humanity are on the largest scale on earth. In this view, Park's bridging role is really a matter of carrying the US message to China.

A different perspective is that Xi is not hiding China's rise as a great power any more, and by embracing it, he is opening the door to a twenty-first-century conversation, which began at the Obama-Xi summit. The United States should avoid a paternalistic attitude toward South Korea, as if it can be expected to communicate what Washington desires, while basking in its new relations with a global Korea with global values that can be trusted. US behavior proves to South Korea that it is serious and balanced in the way it is handling China as well as North Korea, coordinating closely in the alliance and even in taking the Six-Party Talks seriously, proving that it is striving for multilateralism. Yet, a contrasting view held that South Korea must see regional challenges broadly, as the United States does, recognizing that maritime issues involving China are all linked, and it should avoid a narrow focus on winning China's favor that may produce little payoff as it complicates coordination on regional security.

One image of the Chinese challenge to the alliance illustrated the challenges ahead for the alliance. It sees China having a sense of destiny to replace the United States and the international order associated with it, while insisting on a hierarchical regional order grounded in respect for China's central role. Balancing the biggest player in the history of the world will be a gargantuan undertaking. Chinese demands on South Korea to accept the new order and the way China is managing the relationship between North and South Korea will put enormous pressure on the alliance. It will not only squeeze Seoul between two rival claimants for regional leadership, but also leave it in a tug-of-war between two clashing visions of regional identity: a humiliated past that must be reversed by restoring the old order of East Asia, and universal values that must be advanced by extending what has become known as the international community. While in 2013 many are focused on China's call for a "new type of great power relations," as Washington explores what this means for Beijing, the playing field for determining this is, first of all, diplomacy over North Korea and how South Korea maneuvers between the two, divergent opponents.

An optimistic scenario of an unflinching alliance capable of swaying China was also elucidated at the forum. Instead of the defensive alliance on the front line in the struggle against what many in the 1960s–1970s took as the inevitable sweep of communism, expectations have risen for an offensive alliance able to roll back various challenges to the international community, as South Korea serves as a model country and model ally. In this perspective, it conveys to China the message that East Asia has been on the rise since the 1950s, beginning with Japan and continuing with the four little dragons with China a part of this regional dynamism, which, by embracing the region as it has evolved and pressuring North Korea to end its resistance, will contribute to further dynamism. This is essentially the message that Washington and many in Seoul have expounded to Beijing for the past 40 years. It serves to strengthen the alliance, acting in unison.

A sober response concluded that China has already decided that the existing alignment in the region is not in its interest. It is not inclined to accept the US-ROK alliance as a given or that China's rise is just the latest addition to the regional linkup with the international community begun under US leadership after World War II. Instead, it sees China's rise as a game-changer, resulting in a new regional dynamic in which the US role is fundamentally reduced and South Korea is pressured to turn away from or at least narrow the scope of its alliance. Talk about the alliance's future proceeded in the shadow of a debate about China's aspirations, steering discussion toward triangularity.

The Park-Xi Summit

The forum came days before the Park-Xi summit, for which expectations in Seoul had risen quite high, fueled by encouragement from Beijing. South Koreans were starting to think of their country as no longer a shrimp among whales with little prospect of making a splash, but as a dolphin among whales, having new room to maneuver. After Xi agreed with Obama to refuse to recognize North Korea as a nuclear weapons state, reaffirming China's posture of late 2006, the United States welcomed Park's overtures to broaden cooperation with China on this basis.

Yet, concern exists that Seoul will allow itself to become more economically vulnerable to China, considering its interests as in keeping its distance from US and Japanese missile defense plans. In contrast to those who argue that Park is only rebalancing after Lee's error in not consulting with China and that agreeing to new Six-Party Talks on China's terms as well as to a China-Japan FTA parallel to the KORUS FTA are natural steps for South Korea, others see China setting a trap in order to drive a wedge between allies while serving its peninsula plans.

Park's quest for Chinese understanding and trust building is multi-layered, serving to improve coordination on North Korea, to overcome a growing sense that Seoul is being squeezed between Washington and Beijing without room to maneuver, and to lay the groundwork for a new regional architecture. She has kept stressing to China as well as North Korea that she is not Lee Myung-bak. Projecting an image of understanding Chinese history and culture, even taking the trouble to learn the language quite well, Park seeks to convey an image of genuine friendship. If China pressures North Korea, she wants to reassure it that it will not be bypassed either by talks that Pyongyang will pursue with Seoul or Washington or by a process of the North collapsing as Seoul leads in filling the vacuum. This is not zero-sum thinking, and it comes at a fortuitous time for a trial.

Park's chance for success with China depends on Washington's confidence in her. It helps that US admiration for South Korea has been growing rapidly under Obama. It also helps that the US priority is Iran and urgency centers on putting out fires in Syria, Egypt, and elsewhere in the Middle East. That paves the way to moving from strategic to diplomatic patience with recognition that Seoul's chance to defuse the tensions over North Korea may be better than Washington's. Responding positively also serves the goal of improving the US image in South Korea, where those known as the 386 generation—born in the 1960s and socialized by the democracy movement with its cynical view of US hypocrisy about universal values and opposition toward reconciliation and reunification on the peninsula—continue to exert an impact. In Park Washington has a leader who, as with Lee, recognizes that values draw these two states closer and, also, unlike Roh Moo-hyun, that close alliance ties give South Korea leverage over China at a time when triangular diplomacy is the objective.

The Triangle with Japan

New developments in the US-ROK-Japan triangle provoked a lively discussion. On the whole, it was recognized that the ROK-Japan leg of the triangle is being tested in a more serious way than at any time since normalization in 1965. Major differences appeared in recommendations for how Washington should respond. Some voices called for an active, unprecedented role, emphasizing shared values, a common vision for the future of Asia, and, perhaps, acceptance of a degree of US responsibility for leaving islands under uncertain status in postwar treaties. By not pressuring the two allies, as at the time of relatively good relations and shared understanding of threat in 2010–2011, the United States now faces a much worse atmosphere and is paying a big price in the resulting gap that exists in its rebalancing to Asia. A different viewpoint held that US pressure actually would jeopardize both bilateral alliances. US-South Korean ties are not strong enough to bear its weight, and, in any case, the triangle is not leading to some regional security architecture, particularly in light of South Korean opinion about the relative weight of China and Japan in dealing with North Korea and regional leadership. Americans, taking a broad strategic approach, appear to be much more concerned about the impact from poor relations between Tokyo and Seoul than are South Koreans, who are more narrowly contemplating diplomacy related to North Korea, for which they believe Japan has already lost much of its relevance. This is one of the basic causes of triangular malaise.

Japanese seem to be far more positive about the prospects for bilateral ties than are Koreans. After all, US bases in Okinawa and elsewhere in Japan are vital for South Korean security, North Korea has been Japan's security priority since the 1990s, and the record was positive of expanded defense ties with South Korea until the failure in 2012 of intelligence sharing. As Tokyo welcomes full rebalancing by the United States in Asia and moves to join TPP, cooperation with Seoul in collective self-defense and a regional economic architecture should be facilitated. Seeing Abe as willing to downplay historical issues, boosters of trilateralism with Seoul ask only that Koreans recognize that there is no basis for seeing a rebirth of militarism in Japan or for lumping realist responses to growing threats with revisionist views. It is difficult for them to understand how the Obuchi-Kim Dae-jung summit in 1998, in which Tokyo's remorse was clear as was Seoul's forward-looking posture, cannot be a basis of renewal, as, for instance, in 2015 at the time of the fiftieth anniversary of normalization. This cyclical view of relations defies warnings that a fundamental transformation in the region driven by China is leaving precedents in doubt.

The Japan factor's impact on the US-ROK alliance elicited further interest, albeit far less than the China and North Korea factors. If the 2012–2013 deterioration in relations between Tokyo and Seoul is not just another cyclical swing but a sign of a downward spiral, especially if Abe remains in office and Seoul's preference for Beijing endures, then Washington will have much more difficulty in conducting its regional diplomacy, to the extent that the alliance with Seoul may become troubled.

Triangularity with Japan was raised as a growing US concern, requiring a dual approach. On the one hand, it necessitates that even as Washington offers strong support for Abe's economic reforms, recognizing that they are risky, and fully endorses the modernization of Japan's role as a security player in the region, US officials press Abe hard, if quietly, to leave history to the historians. On the other, it requires similar determination to combine wholehearted enthusiasm for a greater South Korean diplomatic and economic role in East Asia with steely resolve to reach an understanding that Japan's expanded military role is consistent with the security needs of South Korea and should in no way be confused with militarism in the past.

In Lieu of Conclusions

From varied perspectives, panelists warned that the US-ROK alliance is coming under greater pressure than before. Precedents of past decades no longer apply, whether in relations with China, North Korea, or Japan. Not only has the alliance broadened in its regional and global contest, but it has also become more vulnerable in its triangular relationships. Closer ties with Beijing may disrupt Seoul's ties with Washington, as under Roh Moo-hyun, and more troubled ties with Tokyo may have a similar effect. Indeed, given Abe's unpopularity in South Korea and his focus on history issues that lead to an overlapping reaction with China, his presence is ripe for exploitation by the gifted foreign minister Wang Yi, who knows Japan well. In the discussion of Wang Yi, the usual lack of clout of the foreign ministry was noted, although a possibility of a more sophisticated strategy under the Xi leadership was not excluded. Park may counter with a sophisticated strategy as well, having built trust with Obama before meeting with Xi and indicating that she would patiently seek evidence of genuine change in China's policy toward the North.

The Asan Washington Forum's diversity of opinion captured the mood in Washington as well as in Seoul as new leadership teams were following up the series of get-acquainted summits in the first half of 2013. Unlike Asan forums in Seoul, representation from other countries in Asia was sparse. This gathering drew Americans and Koreans together to reflect on the alliance that had drawn them closer over six decades.

A new round of diplomacy appears to be gathering steam. Whether the lessons from past alliance relations are learned will be tested in a context broader than bilateral relations. This, no doubt, is what induced many of the panelists to reflect on triangular frameworks. Putting the triangle with North Korea at the center of the second synopsis of this forum will cast more light on the discussion.

Asan Washington Forum Synopsis Part 2: The Alliance and North Korea

In this second and final part of the synopsis of the June 24–25 discussion, we focus on the triangle with North Korea. The foremost priority of the alliance through its 60 years has been preventing another invasion by North Korea. When the chances of an invasion were perceived as diminishing, new strains buffeted the alliance. Discussion often came back to the theme of changes in North Korean behavior and their possible impact on the alliance. In this synopsis of the discussion most relevant to that theme, we look closely at the range and substance of coverage, reflecting the views of US and Korean speakers, some former officials.

In the nearly three months between the Asan Washington Forum and publication of this synopsis, North Korean foreign policy has shifted away from confrontation and refusal to talk to the South. The two sides agreed both on reopening the Kaesong industrial park and on a reunion of elderly relatives. Yet, the early summer Asan discussions remain relevant. The security situation is essentially as before. US policy to negate North Korea's ability to coerce South Korea and Japan with its growing missile force and nuclear weapons stockpile has not changed. Indeed, in calling for a punitive US mission in response to Syria's use of chemical weapons against its own citizens, US officials indicated that if a red line is not drawn in this case, then North Korea as well as Iran would feel empowered to act with impunity. Japan's sense of urgency about how to deter North Korea as well as China has further deepened, gaining new momentum, with Abe's success in the elections to the Upper House. A recent meeting between US and Chinese defense officials was seen as a good step forward, but also as falling well short of the strategic stability dialogue desired by the US side. Even as Japan-ROK relations remain troubled at the political level, at the working level defense officials working with the US Office of the Secretary of Defense are expanding convergence on issues such as extended deterrence. None of these developments substantially alters the strategic environment seen in June. Even the North-South dialogue, more business-like than before, has achieved no breakthrough that is raising much hope.

Events in the summer of 2013 have demonstrated the polarization of positions in a clearer fashion than appeared the case in June when Xi Jinping was signaling to both Obama and Park in summit meetings and, presumably, to Kim Jong-un by refusing to hold a meeting, his displeasure with the North Korean stance and China's interest in denuclearization. In September, US Special Representative for North Korean Policy Glyn Davies toured the region and made clear that China's interest in restarting the Six-Party Talks without prior North Korean affirmation of denuclearization as a goal is a nonstarter. The alliance is being tested by the alternative strategy of China with the backing of Russia in the shadow of a similar showdown over Syria's August 21 use of chemical weapons. Park's "trust-building" puts the focus on Seoul's policies.

The alliance has experienced its most serious trouble when Seoul and Washington have not agreed on how to respond to Pyongyang. In the late 1970s, Jimmy Carter treated the process of normalization with China as an opportunity to withdraw US troops from South Korea, as if the threat no longer needed to be taken seriously. In the mid-1990s, Bill Clinton was more positive about the Agreed Framework and subsequent status of North Korea than was Kim Yong-sam, straining relations. The situation was reversed in 2001–2008 when George Bush took a harder line toward North Korea than either Kim Dae-jung or Roh Moo-hyun. In each case, one side feared that Pyongyang would detect weakness and take a more aggressive posture, leading to tense talks on how to right the ship of the alliance. In 2013–2014, with US attention on Syria and Iran while the ROK is talking to North Korea and also to China, a gap could again open, but to date the two sides keep coordinating.

One concern in the June dialogue of ambassadors was how the process of managing the alliance was damaged in the 2000s and what lessons were learned that could be applied to the challenges ahead. It was found that when troubles arise support on both sides may slip more rapidly than what anyone expected. To prepare for another troubling period it was deemed important to draw clear lessons, anticipate the challenges, and strengthen what is now called a "comprehensive strategic alliance." One lesson is to heighten sensitivity on both sides to symbols of US-ROK equality or public diplomacy in the age of the Internet.

In order to renew the alliance, what approach to North Korea is suitable and how likely is it to be able to overcome different threat perceptions? There is no challenge more important for the alliance than coordination in dealing with North Korea. Opinions on how to do this varied, reflecting different notions of what the overall policy direction should be. One option that was aired was to adopt a strategy of forcing change in the North. Another was to delay and contain the North. A third was to concentrate on China by searching for a shared vision of the future of the peninsula. Reviewing some differences among these approaches offers us an entry point to appreciate the nature of the extended discussion.

Forcing change is an option raised on the American side to reservations from some of the Koreans. It fills the vacuum in US policy after the battles fought among Bush's aides and what some perceive as the absence of a plan in the Obama administration, recognizing that over time an incident may spark a dangerous escalation, exposing the limits of what was depicted as passivity. The first step is to step up diplomacy to find agreement apart from North Korea on the path to a reunified Korea and how to manage North Korea's military. A second step is to intensify the pressure on human rights, demonizing North Korea in a manner that changes the terms of debate, strengthening the case for absorption as the pathway to reunification. Third is an information campaign to turn the public in North Korea and even the elite against the regime, exposing the disastrous nature of its policies. However angered Kim Jong-un and his administration may get, the fourth step must anticipate that through increased military deterrence the regime will be weakened. The example of late 2010 was cited as proof of how tougher deterrence works effectively. This perspective was aired as one train of thought, but it was not echoed widely in June.

The delay and contain strategy starts with the clear assumption that there can be no diplomatic solution without a new calculus by China. Given its attitude, this must be a gradual process, starting with reaffirmation of the September 2005 Agreed Statement as a signal that denuclearization remains in the forefront. The alliance goal should be to find interim measures to constrain the North's nuclear and missile build-up at a relatively low cost. Clinton and Bush both agreed to interim measures. Now these must extend beyond the Yongbyon reactor with the objective of a full declaration of the North's enrichment program backed by intrusive verification if the North is to receive substantial rewards. Should this prove unacceptable, much more limited steps, including dropping some sanctions, can be exchanged for agreement to stop nuclear and missile tests. Even if the agreement is very likely to fail eventually, the freeze has technical value and buys time for changing China's position on the issue. This is more cautious than forcing change. It depends heavily on China, but indications remain that it is not acceptable to that country.

A variant on this strategy for avoiding another cycle with North Korea is to broaden talks with it on economic reforms, recognizing that a decade may be required for Kim Jong-un to be strong enough and convinced of this need. Avoiding the rut of confrontation and buying time, this approach calls for temporary sanctions if new missile or nuclear tests occur, which will sunset after a period if such behavior ends. In the meantime, what is seen as an intermediate arrangement calls for a freeze in the production of fissile material to ensure that North Korea does not become a nuclear arsenal state, while clarifying the Agreed Framework in a manner that clearly explains what giving up the nuclear weapons would mean and even allowing for a flexible definition of denuclearization. This stance is closer to what China is seeking, although many on the US side as well as South Koreans did not endorse it. It was a recurrent strain of thought but not the dominant one in June.

Views on how to manage North Korea ranged widely, although the participation of many with experience serving in the US government and staying well informed of its realist concerns left idealism largely on the sidelines. There were optimists who repeated the familiar refrain that guaranteeing security to the North Korean regime in a far-reaching manner will enable the United States to open the door to economic reform and achieve consensus with China useful for success in the Six-Party Talks. This is essentially the position of mainstream writers in Beijing and Moscow that Washington is the source of the problem and has the responsibility to resolve it. At the opposite extreme were some purists who see no need to compromise owing to the force of allied pressure, the internal forces of change in North Korea, or the second thoughts of China about North Korea's value as a buffer state. This viewpoint, long associated with American conservatives, was also only occasionally volunteered. Intermediate positions were the most prevalent.

Two intermediate positions were more widely aired, although it was not always easy to delineate the line separating them. One holds that military pressure, registering more in Beijing than in Pyongyang, is the best path forward without excluding the potential of diplomacy and positive incentives as evidence of US reasonableness. In early 2003, late 2006, late 2010, and early 2012–2013, US military exercises or show of force, coordinating with South Korea and moves to bolster the alliance with Japan, caught China's attention. Whenever China has pressured North Korea, it has been in response to such moves. Recent signs of such pressure include closer inspections at the border, closure of bank accounts, critical articles in the press, and contrasts in the receptions given to visiting South Korean and North Korean officials. Relaxing the pressure on China, as some expected after the agreement, if left vague, at the June Obama-Xi summit on the priority of denuclearization is deemed counterproductive without a change in China's strategy as opposed to just in its tactics. This line of argument reflects a close reading of Chinese thinking, skeptical of self-serving arguments, and cognizant of interest groups still defensive of North Korea.

The other intermediate position is what many believe Park Geun-hye is pursuing in her visit to Beijing and trustpolitik. It starts from the principle of zero tolerance for further North Korean provocations and insistence that the North respect previous agreements, including those of September 2005 and February 2007. More than Lee Myung-bak, Park is striving for a path to engagement without losing balance from security concerns. It is a realignment policy, patient about dialogue and encouraging to China even as it reassures the United States, obliging North Korea to agree to the objective of denuclearization from the start. A sequence is assumed of less sensitive issues at the outset, then more serious ones as the "Seoul process" keeps in mind human security discourse aimed at reducing the suffering of the North Korean people. This process seeks to balance the two pillars of US deterrence and Chinese cooperation. At the time of the June plenum, this balancing act seemed more possible than in September, when the Sino-US divide sharpened over North Korea's unwillingness to commit to denuclearization and owing to US steps in support of it.

Park has won the confidence of Washington by leaving no doubt that denuclearization (or at least acceptance of it as the objective with concomitant actions) is the prerequisite for advancing relations with Pyongyang. She was clear on this priority when she met China's leaders in late June, urging cooperation between the two sides in assisting North Korea's economic transformation once this essential condition was met and asking China to help to deliver her message to Kim Jong-un. Park's dual summits with Obama and Xi as well as the Obama-Xi summit, where US support for Park's initiative was affirmed, put her at the center of what has the possibility of becoming a new diplomatic process should the new North Korean leadership relent and China persist in applying pressure to that end. In reality, however, China is positioned to steer this process through its application of both carrots and sticks, but it will not have this chance if Park remains firm on the prerequisite.

Although the United States has not drawn a "red line" beyond which it will not tolerate North Korean behavior, recognizing that to do so would invite the North to venture as close as possible to the edge, there is agreement with South Korea that another attack across the border would cross the line. Certain forms of proliferation would do so as well, leading to moves to deny Kim Jong-un certain of his valued assets. While US-ROK differences on what should trigger retaliatory action could test the alliance, there is every indication at present that the two countries are on the same page. With its advanced technology on uranium enrichment or in some other way, the North could aid Iran, as it has in missile technology, leading, if detected, to repercussions from the United States. Without being attacked, South Korea might be hesitant to support the increased risk. In the showdown over chemical weapons use by Syria, there is also potential for spillover.

The future of the alliance may be tested in ways directly and indirectly involving North Korea. US defense cutbacks could lead to a loss of confidence in US capabilities as well as will, challenging ROK reliance on the alliance and, perhaps, its calculus toward North Korea. Growing ROK dependency on a rising China, not only for trade but for managing the threat from North Korea, could shake the alliance. After all, South Koreans already feel closer to China than Americans do, and the gap may widen as Sino-US relations are changing. Another possible source of tension in the alliance is the cumulative effect of failure in the face of North Korea's nuclear weapons build-up, resulting in quite different conclusions by the two allies, and eventually South Korean acquiescence to North-South talks that reduce the pressure for denuclearization. Finally, the abnormal situation of US-Japan military ties growing stronger, notably if Abe succeeds in revising the Constitution and giving new missions and resources to an army no longer obfuscated as "Self-Defense Forces," and Japan-ROK military tensions rising, could do harm to the alliance. Already, opinion in South Korea does not think US handling of Japan is fair and seeks more US pressure on it. While Washington invokes the danger from Pyongyang as justification for its support of Tokyo's military normalization, many in Seoul are inclined not to make this linkage. Of course, there are also distinctly alliance issues, such as burden-sharing and the civil nuclear cooperation agreement, which might play a role in derailing the alliance, but speakers generally treated them as less likely to become inflammatory in the next years.

There is potential for disagreement also in affirming what the alliance is for in a new era. For the United States another priority has been the defense of Japan. To the extent it is beleaguered by North Korean threats as well as maritime pressure from China, one may wonder if South Korea places similar priority on its defense. For both South Korea and the United States an additional goal has been to turn South Korea into a showcase for economic development, proving its superiority in competition with the

North. Originally, the alliance seemed necessary also in order to persuade Syngman Rhee to accept the armistice agreement rather than to march north. During the Cold War the alliance served as a check against communist expansion as well, gradually becoming a model for support of an ally's integration into a dynamically transforming the Asia-Pacific system. With the rise of China and the reemergence of Russia as an opponent of the existing order, the United States is again attentive to maintaining checks and balances among the major powers. It is less clear, however, that South Korea agrees on the priority of either defending Japan or sustaining the balance of power in today's circumstances.

The ideal of a future alliance sounds promising, but its realization was left in doubt at the forum. Instead of unilateral dependence and security still in the forefront, a more equal relationship is envisioned with South Korea serving as the key partner of the United States in the role of a middle power in regional and global affairs. Although this shift was portrayed as a result of mutual convergence, its main thrust is to give Washington the responsible state it has been seeking in support of universal values, such as human rights, women's rights, and mutual security through an international community. Yet, missing in these aspirations is how to counter China's opposition (and use of North Korea's threat capacity) to this sort of alliance, more in conflict with its goals than with the current form of the alliance. After all, given more than two millennia of living with China, Koreans have learned to keep a low profile, not to become outspoken champions of an agenda that would antagonize its strong neighbor. Another concern is that what might appear to be promising for a dynamic Asia rising as a group in the international community looks more dubious for an Asia troubled by demographic, nationalistic, and maritime rivalry challenges, which limit Seoul's room to maneuver. To realize the ideal, moreover, agreement on how to proceed on reunification is needed, as South Korea's leading role enhances its regional and global standing. Continued strife between the two Koreas and a different Chinese approach limits alliance revitalization. Indeed, if polarization occurs between Beijing and Washington, Seoul will find it difficult not to make a clear choice.

The future was put in starker terms with warnings that either the alliance expands to cover a broad range of diplomatic challenges or contracts, endangering its future. The status quo will not do. While there has been creeping expansion of missions, especially as Seoul has assumed greater international responsibility, global leadership is not easy to coordinate with a rising regional role in the face of Chinese and Japanese caution, and with avoidance of peninsular distractions, given North Korea's brusque reminders. One speaker warned that just as Japan failed to become Asia's Great Britain, so too must it be understood that South Korea will not fill that role for the United States. A different model is needed for conceptualizing what a closer alliance would mean in the regional context.

One interpretation of recent events is that Pyongyang has been belligerent even while it was preparing to change direction in order to secure the best possible deal for preserving the regime. In the past few months it has made a fumbling start to a more diplomatic approach, but it is now more urgent for the allies to reach beyond deterrence in order to coordinate their diplomacy closely. In this context, Washington should make clear its firm support of reunification, as well as of diplomacy to entice the North. One step is to endorse Park's proposal for an exchange of offices in Seoul and Pyongyang to speed denuclearization talks and show the North that it is not so isolated that it has nowhere to turn but China. As a good ally, the United States can support South Korea regaining centrality in talks with North Korea, advocates of this position propose, but left vague is what would be the balance of rewarding the North and demanding a change of direction.

A related argument foresees eventual success in the US rebalancing toward Asia allowing increased leadership by Seoul. This essentially comprehensive US strategy has bipartisan support at home and with progress in TPP is opening the door for South Korea, with its rapid transformation of industry and finance and its strong presence in Chinese markets, to seize the initiative as the United States avoids either decline or retreat and China slows its economic march. With Washington obliged to turn much of its attention to turbulence in the Middle East, as the impression is corrected that rebalancing is predominantly of a military nature, Seoul can show its statesmanship in at least three ways: (1) demonstrating that trustpolitik is not just passively waiting for others to prove trustworthy, but it means taking an active role in narrowing differences, including inviting Japan also to pursue this goal; (2) engaging China in joining in a rules-based regional and global order; and (3) not least important, rebuilding channels to North Korea, anticipating its growing interest.

Pessimists about the prospects for denuclearization called for patience, a response made necessary during the Cold War as South Korea eventually pulled ahead of North Korea in development and afterward too, as use of force proved too dangerous. Loss of patience in the late 1970s had a destabilizing effect on the Korean Peninsula, although a different response in 1993–1994 might have been possible in conditions that cannot be recreated. The only hope is to break the family monopoly on power, since there is no prospect that Kim Jong-un would agree to a formula such as 70 percent good, 30 percent bad, which China used to separate a new regime from Mao's legacy. In the interim, some marginal change may be possible with modest reward in order to show the way forward, but the primary response must be deterrence, including expanded missile defenses, with close alliances. Without expecting a lot from China, hope might center on it playing a role in assisting the rise of other leaders in North Korea, who take economic reform seriously, which would be the starting point for intensified diplomacy to begin to address denuclearization.

Disagreements between pessimists and those grasping for some reason to hope arose in a number of panels. Consistent with the aforementioned argument denying the prospect for reform was the assertion that North Korea is built on a structure of lies and myths, which this third-generation leadership has no space to reconstruct without a collapse in legitimacy. In response, it was argued that in this new period, a combination of nuclear weapons and China's rise makes a new deal possible. Holding up a vision of a grand bargain if it were to abstain from further aggression, even if nuclear weapons remain, could, with a balance of carrots and sticks, calm the situation. Kim Jong-un could turn into a reformist, it was suggested, if the balance was well calibrated. Humanitarian assistance is not the same as "buying the same horse again," while sanctions and trade restrictions could be relaxed or tightened, depending on incremental changes. If North Korea used or transferred nuclear weapons, it would know that the policy response would be regime change. Even if there is doubt that the North would reform, this strategy allows for it, while increasing the chance of Chinese cooperation. For substantial aid, including energy, the North would have to grant access to see its centrifuges dismantled in this approach aimed first at the growth of North Korea's threat potential before tackling some additional objectives.

Some panelists found promise in Park Geun-hye's strategy, especially in light of China's changing calculus toward North Korea. Arguing that Beijing now considers a nuclear North Korea as its worst nightmare, a South Korean suggested that Xi is recalibrating, which corresponds to Park's initiative to take advantage of this shift. This means that in the coming period Park does not need to see relations with Beijing and Washington as a zero-sum dilemma. With this starting point, South Korea seems to be free to criticize the human rights abuses of North Korea with impunity rather than to have to take care about crossing a line that Beijing would see as threatening its old priority, the North's stability.

A clear-cut position raised by some Americans with experience dealing with Pyongyang is to engage in no discussions with it that it could be construed as confirming it as a nuclear power. In this situation, the preferred option is for the alliance partners to make clear to China the costs of its enabling assistance to North Korea. This happened in 2010–2011, not least of all in the upgrading of ROK-Japan relations. In the past year that message to China has been garbled. While some discount the impact with the argument that, at last, Beijing is pressuring Pyongyang, advocates of coordinated pressure to send the right signal to both Beijing and Pyongyang worry that Park's leaning to China and away from Japan sends the wrong message. They would wait and see China's actual intentions.

Much of the discussion centered on the North Korea factor and alliance prospects. One dimension is the gap in public opinion toward North Korea and its degree of threat. The two allies have different emotional investment in North Korea and see threat in different ways. As seen in South Korean responses to the inter-Korean summit of June 2000, the alliance can be put in jeopardy by reduced threat perceptions. After all, there has been a tendency to see the alliance only as a counterweight to North Korea. A breakdown of South Korean thinking suggested that one-third firmly support the alliance, one-third are at least dubious about it and amenable to seeing a tradeoff if North Korean ties improve, and the remaining one-third could be swayed. Right now both the government and public in South Korea have a relatively favorable view of the alliance and on synchronization of policies toward North Korea. The US image is not of a power keeping the two Koreas apart. It is not pressuring Seoul to change course. There is no Roh Moo-hyun optimistic about Pyongyang and blaming Washington for North Korea's nuclear weapons and its refusal to reach a deal in multilateral negotiations. Few South Koreans view North Korea as a victim, and, correspondingly, criticism of US policy has diminished. Yet, that could change if Pyongyang shifts course or China was trusted for taking a different approach.

Forward-looking discussions explored ways to strengthen the alliance not just by settling existing bilateral problems, but by deepening ties while giving substance to the goal of a "strategic alliance." Several favorable conditions were noted. First, Park and Obama are admired leaders in each other's country, complementing the positive mutual images of each other strategically, culturally, and economically. Second, partnering in multiple settings already has built considerable momentum. In economic ties, the financial crisis coupled with completion of the Korea-US Free Trade Agreement has spurred closer coordination, as in the G-20. As China's economy enters uncertain times and there is some sign of resurgence in the United States, a successful outcome for TPP with South Korea joining a little later may give a boost to relations. Third, North Korea's threat is unlikely to diminish to the point that South Koreans would reconsider this anchor for the alliance. Yet, such forces were not considered to be sufficient to give a big boost to the alliance. Suggestions were made for a new stage of "rebalancing" toward Asia, in which Obama would articulate a regional strategy, tough on North Korea if it continues on the current path, and dualistic toward China, open to greater cooperation but conditional on its conduct, especially in the face of the North's provocations. North Korea would continue to boost the alliance.

At the root of different interpretations offered on the various panels were three distinct futures for South Korea, the alliance, and management of North Korea. One viewpoint pictures an Asia-Pacific region, in which South Korea is the anchor (or coanchor with Japan) for the US presence. The security alliance remains solid, universal values stand in the forefront in line with South Korea's image as the "leading democracy in Asia," and a shared desire for high-standard FTAs promotes economic understanding. Persuading China to look favorably on the international community and to pressure North Korea to cooperate would be a cornerstone of bilateral ties much stronger than Sino-ROK ones. A second viewpoint envisions South Korea in the role of a bridge between two communities in the Asia-Pacific and East Asia. The Sino-ROK-US triangle would be more equilateral. In addition to supporting universal values, South Korea would show understanding for some values espoused by China for a regional community. It would see itself active in an idealistic quest for regional peace building and as having gained more equality with the United States. Ties with North Korea would improve on terms seen skeptically in the United States, weakening the alliance. A third viewpoint places South Korea as a driver in East Asian regionalism, requiring more cooperation in the Sino-ROK-North Korea triangle and more trust by China. This thinking was least represented at the conference. Rather than a regional leader, the two options in the forefront were South Korea as the leading outpost of the international community or as its bridge to a China-led regional community. Although much depends on China and on Sino-US relations, a major thrust of the two-day discussion was that much also depends on North Korea and ROK-US coordination in dealing with it not only in the face of aggressiveness, but also at times of intensified diplomatic maneuvering, when Beijing is urging Seoul to change direction.

CHAPTER 2

Retrospect and Prospect of the ROK-US Alliance at 60 and Beyond

Choi Kang

The year 2013 marks the sixtieth anniversary of the ROK-US alliance. The catch phrase "We go together," which is used in the Combined Forces Command (CFC), well captures its founding spirit. Since the beginning of the alliance, with the signing of the ROK-US Mutual Defense Treaty in 1953, it has been a key element in both countries' security strategy. For the United States, the alliance has been regarded as a pillar of its East Asian strategy. And, for the Republic of Korea (ROK), this relationship has been a cornerstone of its national security, in the face of North Korea's military threat. For the past six decades, the ROK and the United States have overcome various challenges together, deepening cooperation, not only in the military and security arena but also on political and economic matters. This chapter does not celebrate these accomplishments, but provides background for the Topics of the Month statements aimed at identifying what is needed to bolster the alliance in changing regional circumstances.

During the Roh Moo-hyun period, the ROK-US alliance experienced severe stress, forcing it to adapt to difficult challenges. Each side had its own rationale for the changes, but failed to understand what was motivating the other side. South Korea approached the alliance from a peninsula and subregional perspective with little understanding of the US strategic shift in defense and diplomacy, whereas the United

States approached the alliance rather inattentively to some factors driving not just Roh, but many South Koreans to consider strategy toward surrounding areas anew. Furthermore, the process was poorly managed and trust between the allies eroded to the point that some wondered if the alliance had entered its terminal phase.

Despite the concerns raised throughout the Roh era from both sides of the Pacific, ironically, the ROK and the United States resolved many sensitive and even overdue issues, such as the relocation of US forces in Korea (USFK) and Land Partnership Plan (LPP), strategic flexibility, transfer of ten special missions, and operational control (OPCON) transfer. Unfortunately, these adjustments were carried out without a clear common vision for the alliance. There rarely were in-depth, strategic discussions, which should have guided the entire process. In other words, the many adjustments that were made were driven by ad hoc responses to a succession of pending issues and by the current domestic political mood, especially in South Korea.

With the inauguration of Lee Myung-bak in February 2008, South Korea reemphasized the alliance with the overall goal of transforming it into a "twenty-first-century strategic alliance." For the United States, this development was viewed as a significant opportunity for repairing and strengthening ROK-US relations, as first the Bush administration and then the Obama administration emphasized the implementation of existing agreements, while underscoring the global nature of the alliance. More than before, it had high expectations for the ROK-US alliance.

On June 16, 2009, at their second summit, Lee and Obama adopted the long-awaited "Joint Vision for the Alliance of the Republic of Korea and the United States of America." In it, the two leaders defined the alliance of the future "We will build a comprehensive strategic alliance of bilateral, regional and global scope, based on common values and mutual trust. Together, we will work shoulder-to-shoulder to tackle challenges facing both our nations on behalf of the next generation." During Obama's November 18–19 visit to Seoul, they agreed to hold the US and ROK foreign and defense ministers (2 + 2) meeting in 2010 and to establish guidelines for implementation of the Joint Vision Statement. The joint commitment to the alliance relationship was reaffirmed in the first summit between Park Geun-hye and Obama on May 7, 2013, by adopting the Joint Declaration with the following words:

> We are pleased with the significant progress made in realizing the 2009 Joint Vision for the Alliance of the United States of America and the Republic of Korea, which lays out a blueprint for the future development

of our strategic Alliance. We pledge to continue to build a better and more secure future for all Korean people, working on the basis of the Joint Vision to foster enduring peace and stability on the Korean Peninsula and its peaceful reunification based on the principles of denuclearization, democracy and a free market economy.[1]

Realization of the vision requires much more attention to and clear understanding of the challenges ahead. A clear concept and well-defined nature of the alliance to meet those challenges are also necessary. Attaining a strategic alliance for the twenty-first century should be built on concrete terms and actions since we have already passed the stage of rhetoric and declarations. For that purpose, it is required for us to review the fundamentals of the alliance and to set a new course and roadmap for the alliance in the twenty-first century.

Alliance Adjustment in Retrospect

What We Have Accomplished

Since 2003, South Korea and the United States have discussed and settled various issues related to alliance transformation, using diverse channels. Despite different perspectives, they hammered out agreements on the overall adjustment of the military base system, most notably the relocation of the Yongsan base, the reduction and realignment of US forces, notably the Second Infantry Division,[2] and the overhaul of the "Land Partnership Plan" through the Future of the ROK-US Alliance Policy Initiative (FOTA) Talks, which occurred 12 times before ending in September 2004. They also came to a consensus on issues related to alliance operations such as "strategic flexibility." The Comprehensive Security Assessment (CSA), the Joint Vision Study (JVS), OPCON transfer, and the Command Relations Study (CRS) were discussed at Security Policy Initiative (SPI) Talks. However, differences of opinion came to light in the process, and with regard to strategic flexibility and the CSA, the two countries managed to put some closure on them rather than reaching a complete settlement. Because the process was poorly managed, trust in each other was damaged with the spread of anti-American sentiment in South Korea and the erosion of pro-South Korean, or proalliance, sentiment in the United States.

In the Roh years Seoul and Washington put closure on some items without being able to reach full agreement and with regard to some matters on which they agreed, they put off implementation. More ominously, the two were not able to bridge their differences of opinion over important subjects. The relocation of USFK bases was delayed mainly due to land purchases, environmental cleanup, and cost-sharing for base and facility construction.[3] Regarding the CSA, there were many areas where views did not converge. The two sides closed their talks on USFK strategic flexibility by making note of each party's position, hence going back to square one after earlier agreeing on this. Minister Ban and Secretary Rice worded their outcome as follows:

> The ROK, as an ally, fully understands the rationale for the transformation of the US global military strategy and respects the necessity for the strategic flexibility of US forces in the ROK. In the implementation of strategic flexibility, the US respects the ROK position that it shall not be involved in a regional conflict in Northeast Asia against the will of the Korean people.

The United States asked South Korea to foot a larger share of defense costs on the grounds of "equitable sharing," but South Korea expressed reservations. The transformation of "Concept Plan 5029 (CONPLAN 5029)" to prepare for a contingency in North Korea—into an operational plan—fell through, with differences over the plan's character, its sensitivity, the basic course for responding, and the trigger for the plan's enforcement. South Korea and the United States originally agreed to dismantle the Combined Forces Command (CFC) following the wartime OPCON transfer agreement of April 17, 2012, and create and operate an "Alliance Military Coordination Center (AMCC)." In the course of reviewing this plan, however, South Korea called for an arrangement similar to the CFC, while the United States preferred cooperation among operational units.

With Lee's inauguration, on both sides, expectations soared for restoration of the traditional alliance and transformation of it into a global strategic alliance. The Lee government singled out the "creative development of ROK-US relations" as one of the top ten tasks for arriving at the national signpost of "Global Korea," shorthand for reaching the status of a mature actor in the international community. One of the basic goals was to develop a "strategic alliance" that contributes to the peace of not only the Korean Peninsula and Northeast Asia but also of the world. This was to be done by reinstating traditional ROK-US relations, sharing liberal democratic and market economic values, and further broadening mutual trust and expanding strategic cooperation across a number of political, economic, and social fields. Essentially, for the alliance, three words were highlighted: trust, values, and peace.

Across the Pacific, the US administrations, Congress, experts, and the media enthusiastically welcomed Lee. To strengthen the alliance, Bush and then Obama took positive measures such as freeze of the force level of USFK at 28,500, an upgrade in the ROK FMS (foreign military sale) status, and new measures for extended deterrence. Enhanced cooperation was for combating climate change, increasing energy security, supporting counterterrorism and nonproliferation, peace-keeping operations, and, of course, economic ties through the KORUS FTA. Beyond specific measures was the announcement of the Joint Vision Statement on June 16, 2009, advocated by many security specialists in both countries. It laid out the basis of the alliance of the future (common values and ideals), expanded the arena of cooperation (from security to political, economic, social, and cultural areas), and identified the issues of cooperation at a succession of levels (peninsula, regional, and global).

What We Have Learned

A common problem of both the proalliance and proself-reliance schools in South Korea was a lack of understanding of the essential nature of the transformation of international politics in the twenty-first century and an inclination to focus on domestic elements in interpreting problems. As a result, constantly changing public opinion on specific issues overshadowed the discussion on the alliance. In analyzing problems, it was very hard to overcome a dichotomous framework of conservatives versus progressives and pro-American versus anti-American. Such distractions and divisions hindered understanding of the reasons for transforming the alliance, without which, it was difficult to lay out a suitable direction for the alliance transformation.

To cope with the ever-increasing asymmetrical threats in the twenty-first century, the United States began to implement the concept of an advanced flexible force, which is capable of responding to threats at any time and any place with high mobility and precision strike capability. The shift in the strategic paradigm was accompanied by changes in the traditional alliance system, which was characterized by country-rigid regional defense. Geographical boundaries of the alliance mission lost validity. The US strategy of transformation was neither fully nor correctly understood by the ROK government. Debate over the alliance was mainly driven by fear of abandonment versus fear of entrapment, or proalliance versus proself-reliance. On the one hand, conservatives argued that Roh's mismanagement of the ROK-US relationship and his progressive orientation weakened the ROK security posture. On the other hand, progressives argued for sovereignty, national pride, and a self-reliant defense posture. Neither understood the US transformation strategy at the global level and its implications for the USFK and the alliance. Their arguments were the traditional twentieth-century peninsula-specific ones. Consequently, the domestic debate was counterproductive, failing to provide proper direction for alliance transformation, as the spotlight skipped from one concern of the time to another.

Another lesson we can draw from the past is the importance of cognitive elements in maintaining and strengthening the alliance. Despite all the agreements, the alliance suffered from an erosion of trust and confidence, which was rooted in a perceptual gap that did not become the object of serious effort to overcome it. Contrary to the official statements and explanations of both the ROK and US governments throughout Roh's years, friction, even conflict, between the policymaking as well as opinion-shaping groups of both countries went beyond simple policy differences. They became suspicious of each other's intentions, sincerity, and integrity. This resulted in misinterpretations and left a huge emotional scar.

For example, the Roh and Bush administrations had quite different outlooks on North Korea, China, and the desired regional security architecture. Seoul thought that the latter exaggerated the North Korean threat and did not understand its intentions due to its preoccupation with the war on terror, whereas Washington believed that the former was too sympathetic to North Korea and underestimated the gravity of its threats. They also differed in estimating the long-term regional security environment, most notably China's implications for regional peace and security. Such cognitive differences exerted great influence over policy coordination between the two allies. While they were able to hammer out narrow agreements on pending issues, the ROK and the United States were becoming less trustful of each other's genuine intentions.

One more lesson follows: the importance of a vision for the ROK-US alliance in showing the way to issue-driven adjustments. The Roh and Bush administrations failed to identify the rationale of the future alliance or its preferred shape, in contrast to what was achieved for the US-Japanese alliance during the 1990s. The adjustment process between South Korea and the United States failed to identify roles, missions, capabilities, strategies, and the structure of the evolving alliance. It dealt with the issues. Even if the United States had an idea of where it wanted the alliance to go, the ROK did not have much more than the desire to be less dependent—"self-reliant defense." The adjustment process itself became a source of misunderstanding.

In sum, a lack of understanding of the transformation in international relations, a lack of effort to narrow the perceptual gap as trust eroded between the two allies, and the absence of a vision are much more important than the differences over specific, outstanding problems at hand. To put it differently, the Roh and Bush administrations did not have any in-depth strategic dialogue or coordination. Rather, they opted for tactical discussions. Thus, they failed to diagnose the symptoms and to identify the fundamental causes of problems. These experiences should be seriously taken into account in realizing the ROK-US strategic alliance in the twenty-first century. The past guides the future.

Alliance Prospects: Future Challenges and Tasks

After Lee's inauguration, a series of meaningful agreements were adopted and implemented. Among these, Joint Vision for the Alliance on June 16, 2009, was significant in setting the direction for the alliance. It was endorsed and underscored by the Park and Obama administrations through the Joint Declaration in May 2013. To realize these aspirations for a twenty-first-century strategic alliance, many things must be done.

Defining "Strategic" Alliance

The first task is to clearly define and operationalize the strategic alliance. It is necessary to ask "what is the difference between alliance and 'strategic' alliance?" This involves clarifying the roles and mission of the alliance in the coming years. An alliance is formed to cope with a common threat(s), notably military in nature, as a way to make up for a deficiency. It is different from other forms of relations because it is primarily based on close military coordination. An alliance can be defensive and passive in nature; promoting stability by a balance of power, or maintenance of the status quo, while reacting to any action intended to weaken one or both of the partners. Such has been the case in geographically bound alliances such as, the postwar NATO, US-Japan, ROK-US, US-Australia, and US-Philippines alliances.

Both the Lee and Park administrations identified three elements of a strategic alliance of the twenty-first century: values, trust, and peace. The two allies share common values; their relationship is based on mutual trust; and the objective of the alliance is peace. By finding new meaning in these concepts, they propose that the ROK-US strategic alliance become different from the traditional, threat-based alliance. To do this, both sides emphasize expanding the scope of the alliance. Geographically, the alliance is no longer confined to the peninsula. Rather it is intended to cope with regional and global security issues as well. Functionally, the alliance covers not only military issues but also nonmilitary ones, especially since 2009, in the shift to comprehensive cooperation. The Joint Vision of June 2009 reads as follows:

> [O]ur security alliance has strengthened and our partnership has widened to encompass political, economic, social and cultural cooperation. Together, on this solid foundation, we will build a comprehensive strategic alliance of bilateral, regional and global scope, based on common values and mutual trust. Together, we will work shoulder-to-shoulder to tackle challenges facing both our nations on behalf of the next generation.[4]

In the Joint Declaration, after addressing their concerns over North Korea, both sides emphasized the cooperation between the two countries for regional as well as global security and economic growth by stating the following:

> Based on the solid US-ROK alliance, we are prepared to address our common challenges and seek ways to build an era of peace and cooperation in Northeast Asia... We will strengthen our efforts to address global challenges such as climate change and to promote clean energy, energy security, human rights, humanitarian assistance, development assistance cooperation, counter-terrorism, peaceful use of nuclear energy, nuclear safety, non-proliferation, cyber-security, and counter-piracy.[5]

The "strategic alliance" should be geared to shape the future, not just respond to situations. Nowadays the function of an alliance is not just for maintaining the status quo, but for actively developing the desired strategic landscape. For that purpose, proactive and preventive measures are desired, and a combination of means across fields is going to be a dominant feature. The concept of DIME (diplomatic, information/intelligence, military, and economic) has become very popular in the discussion between allies. Comprehensive intergovernmental cooperation, which is far beyond just mil-to-mil cooperation, is necessary, more than ever.

Responsibility sharing, or benefit-and-burden sharing, is one of the key elements. Alliances used to be a kind of patron-client relationship. Mutuality is now being emphasized on both sides. One's strategic value is increasingly determined by how much one contributes to the attainment of common goals. The issue is how to define and share responsibilities. In other words, who takes the leading role, while the other assumes the supporting role? It is necessary to think of the relative weight of the national interests at stake for each country, and at their relative capabilities, that is, who has what and how much.

Based upon the aforementioned criteria, we can say that a strategic alliance is proactive, balanced responsibility sharing, issued-based, adaptive, future-oriented, and a comprehensively integrated alliance. But the core of a strategic alliance is the enrichment of military coordination by deepening cooperation. This actually encompasses some elements of a regional alliance, but goes beyond simple geographical extension of the ROK-US alliance. It can be called a "multidimensional comprehensive alliance." With this new framework, South Korea and the United States should identify and deter the potential threats to peace and security in the region and the world. A new alliance should not be targeted toward a specific state or bloc. Instead it should be a tool for realizing cooperative security.

Strengthening the Fundamentals

There are several key elements in realizing a comprehensive, multilayered, strategic alliance. These elements can be considered as the fundamentals of the alliance. The future of the strategic alliance depends heavily on how much they are consolidated.

First, there should be a common threat perception, or assessment of the future security environment and challenges, between the allies, based on the existence of shared national (security) interests. Such shared interests are a necessary, but not sufficient, condition, since it is possible to choose other means of realizing those common interests than forming an alliance. There should be something else that makes an alliance different from other relationships in international relations. Of course, threat perceptions themselves evolve as changes in the internal and external environment are perpetual. It is not difficult to find cases where allies have different threat perceptions, and cooperation grows weaker as the relationship is strained. The future of the ROK-US alliance greatly depends on how effectively the two will replace the original rationale for forming the alliance—the North Korean threat—with the threats to common values and interests in the twenty-first century. To strengthen the ROK-US alliance in the twenty-first century, it is necessary to strengthen this foundation.

Second, another important element in alliance relations is trust. The existence of a common threat perception and national interests does not ensure solidarity. These can change due to changes in the internal and external environment. Allies may shift positions. They may fail to carry out the commitments they have made. Under conditions of uncertainty, what ensures the integrity of the alliance is trust, that is, belief in "whatever you do, I am with you." Of course, a common threat perception enables the alliance to endure not only during wartime and contingencies but also during peace-time. But trust that the ally will fully honor its commitments maintains the robustness of the alliance regardless of changes in the internal and external environment. Trust is becoming more important in an era of defense transformation, and it requires fundamental recognition of the congruence of strategic security interests between the allies. Furthermore, it is the basis of common identity in values and vision. To upgrade the ROK-US alliance, efforts should be made to deepen a feeling of "we go together."

Third, there should be institutional mechanism(s), between the allies to support the realization of the cognitive foundation. Otherwise, it would become "No Action Talk Only," and it could not operate effectively. To strengthen mutual trust, it is necessary to have an effective institutional framework, including treaty-bound responsibilities. In introducing such a mechanism, the following points should be seriously taken into account. While honoring the basic spirit of the ROK-US Mutual Defense Treaty, its preamble reads:

> The Parties to this Treaty,
> Desiring to declare publicly and formally their common determination to defend themselves against external armed attack so that no potential aggressor could be under the illusion that either of them stands alone in the Pacific area,
> Desiring further to strengthen their efforts for collective defense for the preservation of peace and security pending the development of a more comprehensive and effective system of regional security in the Pacific area.

Thus, from the beginning the ROK-US alliance has not been exclusively confined to the Korean Peninsula. This has significant implications for the future ROK-US alliance. First of all, the expression "*no potential aggressor could be under the illusion that either of them stands alone in the Pacific area*" implies that the objective of the alliance is not only the deterrence of North Korea's use of force against South Korea but also responses to diverse potential security threats in the Pacific area. The scope of the alliance covers not only the Korean Peninsula but also East Asia and the Asia Pacific, and it can be extended to cover global issues due to the increasing linkage of challenges across regions in the world. Furthermore, it is geared toward the establishment of a *"comprehensive and effective security mechanism"* the cornerstone of collective security in the region. It means that the ROK-US alliance and multilateral security cooperation are not mutually exclusive.

Second, the ROK should be more concerned with the meaning and implications of the word "mutual" in the Mutual Defense Treaty. Such a treaty implies that when one party is in danger, both parties respond together. Until recently, the ROK-US alliance has been usually understood as the US commitment to the defense of the ROK, however, as it has developed its national power over the years, the ROK should recognize the fact that areas for cooperation over the US security concerns have also widened, and that it should be able to share the responsibility. To develop a balanced, robust alliance with the United States, it is necessary for the ROK to assume greater responsibilities and to increase its strategic value by making meaningful and substantial contributions to the US efforts in managing and transforming the regional and global order. At the same time, the United States should also be well aware of subregional level concerns and issues that affect the ROK's national interests directly.

Third, the ROK and the United States should think of ways to utilize the possibility of flexible interpretation of phrases in the Mutual Defense Treaty. Some argue that to extend the scope of the alliance beyond the Korean Peninsula it is necessary to amend the Mutual Defense Treaty. Otherwise they claim it is a violation of the treaty. However, it is unnecessary and counterproductive to amend the treaty. Rather, as the United States and Japan did in 1996 when they made a declaration of new security cooperation, the ROK and United States should set the direction of the alliance within the framework of the existing treaty. In addition, it is possible to lay out a kind of security cooperation guideline or principles and to adjust the current command structure—the Combined Forces Command—accordingly for the sake of ensuring effectiveness in carrying out new roles and missions.

Fourth, even when the ROK and the United States aim at a "comprehensive alliance," the core element should be deepening military cooperation. This makes the alliance different from other types of relations and more important. "Comprehensive cooperation" should not mean weakening the military security commitment to each other. Rather it should be interpreted as a way to strengthen the alliance by adding other elements to the existing core and by constructing a more complex, multilayered relationship, which will ensure the stability of the security commitment and enhance military cooperation. With a complex alliance, the ROK will be able to maintain robust security collaboration, backed up by the expansion of support for its national interests in the political, economic, and social arena.

The key to repairing and strengthening the ROK-US alliance is to consolidate the foundations of the alliance by stepping up efforts to place priority on fine-tuning coordination on each outstanding issue and promoting shared understanding. The United States and South Korea are bound to have incongruent points of view and perceptions, but it is vital that the two countries make efforts to bridge their gaps to the maximum extent and foster trust. In order to move toward a "strategic alliance" that aspires to share values and broaden strategic cooperation, what the two countries need is a set of institutions, and measures of closer consultation capable of ensuring "comprehensive interoperability," where even situational assessments and prospects and means of planning and response are shared.

For its part, the ROK needs to take the approach of expanding pursuit of its national interests through the utilization of its alliance with the United States as a means of counterbalancing the burden of maintaining the alliance. The ROK also needs to take advantage of the alliance to supplement its weaknesses—notably its deterrence against North Korea and balance of power vis-a-vis neighboring countries—and raise its stature in the security realm and regional security structure. Seoul also needs to make an effort to enhance its strategic value by developing its own areas of specialization, which befit its international prestige, national power, and image in the outside world. To this end, the ROK, a successful model of democratization and economic development, may examine ways to blend "soft power" with other positive elements it possesses and make use of it as an asset for the alliance. The ROK can also consider ways to take advantage of the know-how it has accumulated in the fields of peace-building and social reconstruction to establish its own areas of expertise.

Finally, based on the experience during the Roh administration, management and resolution of pending issues is critical to sustaining a robust alliance relationship. Three issues have already begun to get public attention: the ROK-US nuclear cooperation agreement, wartime OPCON transfer, and the Special Measures Agreement. Each has potential to ignite an emotional response from the public and become a burden for the government in managing the alliance. Both sides must approach these issues with a skillful approach and mutual respect.

Conclusion

To realize a strategic alliance, a clear guideline, which goes beyond simple rhetoric, must be adopted and an action plan devised through intense discussions. For that purpose, it is essential for South Korea and the United States to undergo a bottom-up review process of the alliance. It is important to pay careful attention to: identification of challenges the two must cope with together (for what); the division of labor (who should do what); plans and strategy (how); and cooperation mechanisms (through what). Along these lines, it is important to think about how to enrich the contents of military cooperation as the backbone of the alliance. To avoid any misunderstanding and overexpectations of the United States, South Korea must make it clear what it can and cannot do on the basis of undiminished mutual respect.

Notes

1. *Joint Declaration in Commemoration of the 60th Anniversary of the Alliance between the Republic of Korea and the United States of America*, May 7, 2013.
2. In the FOTA process, the United States revealed a plan to reduce the size of USFK by 12,500, a process that continued until early 2008, when Lee and Bush agreed to freeze the USFK at 28,500 at their first summit in April 2008.
3. As of January 2008, 38 out of the 79 USFK bases targeted for relocation had been returned, and of the remaining 41, 9 were supposed to be returned to South Korea that year.
4. *Joint Vision for the Alliance of the Republic of Korea and the United Stated of America*, June 17, 2009.
5. *Joint Declaration in Commemoration of the 60th Anniversary of the Alliance between the Republic of Korea and the United States of America*, May 7, 2013.

CHAPTER 3

Synopsis of the Asan Seminar: "Managing ROK-US Relations"

Gilbert Rozman

As introduced by Choi Kang, vice president of the Asan Institute, this seminar was a follow-up to the June 2013 Washington Forum, which had taken stock of the sixtieth anniversary of the ROK-US alliance. Choi noted that difficult issues are becoming more prominent, raising more uncertainty about how to realize the comprehensive, strategic alliance for the twenty-first century. While the discussion in June was mostly about bilateral issues, this time trilateral matters took center stage, not so much about North Korea or even China as in June, but Japan above all. Questions posed by the audience to Kurt Campbell, who first gave a keynote address and then responded to queries raised by Choi, overwhelmingly centered on ROK-Japan relations. To a surprising extent, so too did questions addressed to Mo Jongryn, who led Session 1 with his presentation on Korea and the Liberal International Order, and to Kim Jiyoon, who led Session 2 with her presentation on Korean public opinion and the ROK-US alliance. Choi has led the Topics of the Month discussion on the ROK-US alliance in the Asan Forum, and it is fitting that we link this synopsis to that discussion, including a new article directly on the impact of ROK-Japan relations by Bong Youngshik as well as a rejoinder by Rust Deming. The October 25 seminar raises the intensity of the exchange, as worry about how to manage the ROK-US alliance is informed by ROK-Japan ties.

The seminar steered discussion toward the shape of future relations, interpreting the concept of a strategic alliance as proactive in preparing for the future and that of a comprehensive alliance as extending beyond defense against North Korea to the region and the globe. Over the period of Lee Myung-bak's leadership, the alliance has been tested on bilateral matters and in provocations from North Korea, and it has emerged stronger than before, enjoying exceptional bipartisan support within the United States and public approval in South Korea. After moments of anti-American emotionalism in South Korea, uncertainty about the KORUS FTA, and doubts about a lack of US focus on East Asia, the alliance held steadfast, and Park Geun-hye began her term as president with strong expectations on both sides of new momentum. If these sentiments prevailed in June, a cloud hung over them at the recent seminar.

There was a worrisome undertone in October not present four months earlier. It surfaced first in vague references to the need for psychological adjustment in both states from a paternalistic state of big brother, little brother in a context of rising nationalist sentiments in South Korea. It also hovered in the background in talk of Global Korea, which the United States strongly welcomes along lines already seen, especially at a time when other states are pulling back in their global aspirations, but when the subject shifted to a new grouping of middle powers from September 2013, known as MIKTA (Mexico, Indonesia, Korea, Turkey, and Australia), pursuing their own agenda, it was less clear what gap between the United States and China they would fill and how that would relate to US global leadership. Also, on the subject of trade, the celebratory mood over the success of the KORUS FTA, whose opportunity has been successfully seized by Korean companies, and optimism for Korea leading a second wave of states into the TPP and reaching beyond past competition in which it and the United States as well as Japan had failed to cooperate much in engaging Southeast Asia was juxtaposed to awareness of the shift from trilateralism as the focus in a CJK FTA to focus on a bilateral CK FTA without Japan. On these themes the mood was largely upbeat, suggesting that open economies remain a positive force in ROK-US relations. It was even suggested that Korea should not set its sights so low as to think that it must be only a middle power when in the emerging division of labor with regard to soft power and innovative industrial leadership it could stand ahead of the pack. It was only when other themes arose that the concerns became much greater.

One theme did center on bilateral relations. It covered OPCON transfer, missile defense, and responses to the revolution in military capabilities that has already greatly increased the alliance's deterrence capacities. There was talk of the urgency of managing these issues carefully, especially in the wake of contradictory remarks about whether Korea is joining the missile defense system being developed by the United States. If these issues are not handled well, it might seem that the United States is withdrawing from Asia or that South Korea is distancing itself from the alliance. Chuck Hagel had spoken of the interoperability of missile defense during his recent visit to Seoul, but Koreans have tried to be cautious in their language, as if to draw a line between theater defense limited to the North Korean threat and more "global" missile defense that China interprets as containing it. While it seems likely that the US side understands the need to convey this sort of strategic message to China, the Chinese side is more suspicious that there is no real distinction. Even in discussing what starts as a bilateral topic, its trilateral ramifications soon appeared.

The trilateral North Korean theme spilled into quadrangular discussion of China. As is often the case, listeners were attentive to any indication of differences between Seoul and Washington in dealing with Pyongyang. From the US side came the view that over the past 20 years every possible diplomatic approach has been tried. Although the door should be kept open for any new prospect to materialize, the best chance for diplomacy now is to maintain robust deterrence. Recent signs that China is exasperated by North Korean provocations especially warrant diplomacy toward it. Indeed, Washington is welcoming Seoul's leading role, rather than Beijing's or its own, in diplomacy related to North Korea, hoping that, in the process, Chinese grasp the sharp contrast with Pyongyang's behavior, trust Seoul more, and are persuaded to work with it more closely. Yet, in the questions raised, there appeared to be an undercurrent of uncertainty about Park's "trustpolitik" toward North Korea, which translates into "trustpolitik" toward China and its handling of its ally, to the degree that some seem to think that rather than gaining China's trust, Park is in danger of being used to strengthen the clout of a country that wants to drive a wedge between Seoul and Washington, to isolate Tokyo, and, eventually, to boost Pyongyang's hand.

Park's "trustpolitik" and "Northeast Asia peace and cooperation initiative" put Seoul at the center, but they leave the regional role of the alliance unclear. Those who recall Roh Moo-hyun's 2005 appeal for his country to become a "balancer" may find some resemblance. In contrast to the deep suspicion that greeted Roh's idea, recent US responses suggest a willingness to listen and to start with soft issues, such as fishing zones, which may some day make possible a Northeast

Asia community. Yet, the bottom line for US support is regarded as so unlikely to be realized that just the fact that hopes are being raised leads to concern. Unlike Beijing or Moscow, it holds that Pyongyang must reaffirm past nuclear agreements as a precondition that must be maintained. Realism demands that memory of past failures with the North must temper expectations that new ideas are within reach to achieve a breakthrough.

In response to two of the panels there was debate about what it means for South Korea to be a middle power. Lee Myung-bak was known to link this notion to Global Korea, putting Seoul in the middle between developed and developing countries with a development agenda, leadership in transnational summits, and lessons from its recent transitional experience. This was very different from Roh's thinking and served to bolster the alliance. Park is less focused on Global Korea, raising questions about whether she has revived the idea of middle power diplomacy between what is increasingly seen as a G-2. Other countries, such as Australia, have close economic ties to China, but do not elevate themselves into the strategic middle or harbor real aspirations of mediating between the two great powers. On the US side concern was evident that the alliance could be damaged by old balance of power thinking. Many appeared to doubt that Park has the leeway to exercise real leverage on either China or the United States, which, in a more polarized environment, may tug in opposite directions on South Korea with other countries having little impact on the triangle.

What began as a few exchanges on a middle power role in the first panel turned into an extensive assessment based on Jongryn Mo's presentation. He noted that middle powers have largely been absent from debates about international relations, but in a polarized Sino-US divide, they can set an example for developing countries, fill a gap in mediating between the G-7 and the BRICS, and win widespread trust, in part due to the absence of an imperialist past. Citing his coedited book with John Ikenberry, *Rise of Korean Leadership: Emerging Powers and Liberal International Order*, Mo gave many examples of Seoul setting an example and taking the initiative, but he was careful to explain that this has mainly been in the realm of intellectual roles, not hardware (such as a substantial level of ODA), and he expressed concern that Park is not as committed as Lee was to Global Korea and its intellectual leadership impact.

Mo raised some questions about his middle power thesis, and these were followed by questions from Sheila Smith and Victor Cha, who served as discussants, and then from the audience. Will South Korean domestic politics sustain this leadership role when the public elsewhere is pulling back from it, in part because progressives, who are more supportive, are losing power? Is there evidence that the middle powers are seriously cooperating? Are the great powers leaving an opening for this kind of role, or is polarization intensifying with limiting effects? Were special circumstances in 2008–2012, which are unlikely to be repeated, conducive to a greater role for South Korea? For instance, ROK-US ties were particularly close, North Korea was isolated more than in other periods, China overreached and left other states seeking more support, and the global financial crisis and weakened US image created an opening. Sheila Smith emphasized the strong US encouragement and South Korea's success in drawing lessons from its own financial crisis a decade earlier, but she wondered if the South's enhanced role is sustainable or whether it has a strategy to make it so. She wondered if the alliance has been positive or negative for this enhanced global role, whether Seoul has succeeded in tempering rivalries, and what are the trade-offs between its global and its regional role. Victor Cha was also positive on South Korea's role as an intellectual leader and on its capacity to empathize, as a country that had passed through one state and entered another, for example, to a major peace corps source, an ODA donor, a developed state, and a democracy. He saw potential for new roles. In the 123 talks it could help to set norms for nonproliferation and nuclear safety. In this example, as in the Six-Party Talks when it, along with the United States, offered good ideas, South Korean leadership would require thinking principally of what serves international peace and stability, not of narrow national concerns.

Kim Jiyoon in the final panel drew attention to new findings on public opinion in South Korea. She started by noting its substantial impact, as in the 2002 election of Roh on a wave of anti-Americanism and the 2012 last-minute decision by Lee not to sign an intelligence sharing agreement with Japan. Three main findings this fall are: (1) increased support for conservatives, led by the youngest and oldest voters, as progressive support has fallen sharply; (2) even stronger support for the US alliance among all age groups; and (3) China's much improved image after the June Park-Xi summit. Even after reunification, 80 percent of respondents say the alliance would be necessary. New perceptions of China's reactions to North Korea have helped its image, which has benefited by a drop from 61 to 47 percent who consider it to be the biggest threat to reunification. There are many more interesting results in the detailed presentation by Kim Jiyoon, but her work can be examined separately, and it will be presented in an article planned for the fourth issue of this journal.

Elaboration on the public opinion survey came from Woo Jung-Yeop, who asked if respondents saw the alliance as directed against China once reunification had been achieved and suggested that about half did, but, on the whole, US efforts to proceed in this direction would strain the alliance. Already, the improved image of China, he argued, was testing the alliance. When South Koreans had perceived China siding with North Korea, their image of China had deteriorated, but in the summer of 2013 this changed. This finding points to the volatility of China's image with uncertain impact on the US image. If earlier there were indications of an inverse relationship between the images of these two great powers, the fall 2013 survey showed no such thing. Is this a temporary aberration, many were probably asking, but the discussions spent little time on China, given a preoccupation throughout the day with ROK-Japan ties.

ROK-Japan Relations and Their Impact on the ROK-US Alliance

Reversing the order of coverage above and in the day's agenda, this synopsis starts with public opinion data and discussion. Kim Jiyoon's shocking finding says that, in country favorability, the United States stands at 6.4, China at 4.2, and Japan at 2.4, lower even than North Korea's 2.7. The older the cohort, the worse the rating for Japan, with a high of 2.8 for those in their twenties and a low of 2.0 for those in their sixties. In choosing one country as the biggest threat to Korean unification, older cohorts also were most negative about Japan, putting it above China although those in their twenties were more than twice as likely to select China, which overall fared worse than Japan as a threat. Younger respondents were more fixated on a values gap with China over its political system and human rights, while the history values gap with Japan is foremost in the minds of the oldest respondents. These and other findings regarding Japan elicited many questions from the discussants and the audience.

Woo Jung-Yeop noted that after half a century of living with an alliance that paled before the closeness of the US-Japan alliance, in 2009–2012 South Koreans basked in the feeling that now their special relationship with the common ally was closer and had a broader global significance. Then, abruptly in the last few months, they awoke to the shock of the strengthening of US-Japan military ties, at a time they have come to see Japan's military as a threat linked to historical revisionism. Questions were informed not just by the reported public opinion data, but also by familiarity with the South Korean media coverage of Chuck Hagel's back-to-back visits in October to Seoul and Tokyo, which is reflected in Bong's Chapter 5. Some detect a palpable sense of alarm in South Korea over US support for Japan's military build-up and moves toward collective defense. Japan is now casting a shadow on the alliance.

Gordon Flake and Bruce Klinger were the two discussants for Kim Jiyoon's remarks. Flake remarked on the changing image of a progressive as pro-North Korea at a time of the North's belligerence, which reverberates in images of a conservative and of the US alliance, while he took particular note of the drop in Japan's rating from 4.2 in 2010 to 2.7 in 2013 with the biggest drop in the past year, finding it an indication of paying attention to Abe's words rather than to his actions. He found it perplexing that Japan should be lower than North Korea, suggesting a parallel with some 2006 negative findings on the United States, rooted in the idea it might start a war with North Korea. Disagreeing with those who see the response to Japan as a bottom-up phenomenon, he attributed the problem more to the leadership than to the public.

Japan emerged as a topic in the panel led by Mo Jongryn too. With Soeya Yoshihide, well known for his interpretations of Japan as a middle power, in the audience, it was only natural that thoughts would turn to parallels with South Korea and the prospect of joint endeavors by two middle powers situated next to each other. For some in the audience, Mo's assertions about South Korea's potential impact were a matter of deja vu, familiar from arguments about Japan's potential role two–three decades earlier. Japan too had hoped to become a cultural force noted for its lifestyle and soft power. It envisioned itself as a bridge to developing countries, offering ODA while standing as a model outside the West for rapid transition into a developed state. Mo spoke of a bureaucratic model of leadership toward multilateralism, which Japan championed when public support was of little consequence and South Korea now may be trying to go beyond, in ways that proved difficult for Japan. Sheila Smith was attentive to this parallel, pointing to Japan's disappointment as its global role failed to meet expectations, public support did not materialize, and its soft power slipped as its economic influence was falling. Multilateralism independent of US leadership proved elusive in Japan's case, especially in the face of China's rise, and the same factors may doom South Korean ambitions. Mo's comment that South Korea has no hegemonic past along with evidence that its citizens are more international in their outlook, as seen in study abroad, casts some doubt on the analogy. Victor Cha noted, however, that Koreans need to be ready for the disrespect that comes with playing an active global role, as both the United States and Japan have experienced.

After stressing that Japan was 20 plus years ahead of South Korea in facing similar challenges, Soeya went further in arguing that by ignoring its role as a middle power Koreans are failing to draw the appropriate conclusions for their own country. After all, he observed, the two states vote in the United Nations in a similar manner, their ties to the United States are similar, and academic theory has difficulty separating the two. Why, then, do South Koreans ignore Japan as a partner in its quest for more middle power clout and overlook the merits of sticking to the bureaucratic model without arousing the public in a way that undermines that model, in the process, undercutting the prospects for a sustainable, global middle power strategy? The view that Japan is South Korea's logical partner was repeated by others, along with doubts that its middle power idealism is, in the long run, compatible with a close US alliance, since the goals appears to be a bridge in a power balance, driven by national identity more than by national interest. From the Korean side one heard a contrasting view that to the extent that the United States is opposing South Korea's middle power strategy it is stifling its ally, especially by casting doubt on Park's trademark themes, such as the Northeast Asian peace and cooperation initiative, but others held out more hope for Park's cautious invocation of the middle power theme and appeal to nationalism. Her approach to Japan, however, defied this optimism.

The discussion on October 25 came against the backdrop of divergent US and South Korean reactions to the successive visits of Chuck Hagel to Seoul and Tokyo. Anger felt in Seoul toward US support for Japan's "remilitarization" contrasts with loss of patience in Washington as well as in Tokyo at what many regard as "wild" Korean statements on Japan as if it is mired in its pre-1945 past. In the summer of 2013 the focus of US concern with national identity run amok was Japan. That fall it appears to have shifted to South Korea, although concern over Abe remains. Given divisions in views toward Xi Jinping as well as Abe, we may be on the verge of a serious test in ROK-US relations in the coming months. This was the mounting concern as the US-ROK-Japan triangle overwhelmed the bilateral focus as the seminar proceeded.

The urgent first step for extricating the ROK-US alliance from the trouble it is facing is for Koreans to stop making collective defense in the Japan-US alliance a litmus test for trust in the United States. The media has been doing so; although public opinion toward the United States earlier this fall was remarkably positive. When a Korean journalist asked if Washington had not decided to side with Japan against Korea by supporting Japan's embrace of collective defense and if Hagel's statements in Tokyo following Park's warnings about Japan were not an indication of defiance of South Korea with the potential, presumably, to arouse anti-Americanism, many on the US side may have been awakened to a threat to mutual trust that few had anticipated.

Recognizing the seriousness of the problem, Kurt Campbell firmly expressed the US position that Japan is a peace-loving country, which is only responding to challenges of today, not influenced in its defense posture by thinking of the 1930s. It is a loyal US ally with modest defense expenditures and public opinion with peaceful ideals. He was pressed to explain what Washington could do in addressing the deepening ROK-Japan divide and the growing anxiety in South Korea over Japan. One person suggested that if Obama had been able to make his planned trip to Southeast Asia he could have arranged a three-way meeting with Park and Abe. Another proposed that US admission of responsibility for mishandling the history and territorial issues in the 1950s would assuage hurt feelings, helping ROK-Japan relations. Campbell made a more modest suggestion that the United States create a framework for discussion, praising Hagel and Kerry for their efforts, but he warned against pushing others into a corner, turning the focus into how it is the problem. It would be unwise if officials put Obama in a position with an unpredictable outcome. Delving further into history only draws more attention to national identities, worsening the situation. A more promising US role is to refocus the discussion on the way forward. The very fact that Abe has opened the door to more historical attentiveness and Park has decided to put history first has left Washington in a perilous position, many likely concluded. At a time when the bilateral alliances and also trilateralism may be most essential, the latest developments in ROK-Japan relations and their spillover into South Korea's relations with the United States are stirring anxiety that is entirely unwelcome, if we take the response at the October 25 seminar as an indication of the mood in Washington.

CHAPTER 4

Synopsis of the Asan Institute-Pacific Forum CSIS Conference on US-Japan-ROK Trilateral Relations

J. Berkshire Miller

In July 2014, the Pacific Forum CSIS and the Asan Institute cohosted a conference in Maui, Hawaii, on trilateral relations of the United States, Japan, and the ROK. The larger focus was on planning for a contingency on the Korean Peninsula that would involve trilateral cooperation. While all sides appeared to agree on the importance of working in unison, especially with regard to deterring and responding to North Korean provocations, doubts prevailed on the prospects for the political flexibility needed to enhance trilateral ties, as a result of the strained relationship between Tokyo and Seoul.

The importance of enhanced understanding among the three on a number of critical issues, including Japan's recent national security and defense reforms, was discussed as candid views were exchanged. In particular, there was some fruitful dialogue on the Abe administration's recent Cabinet decision to reinterpret its constitutional right to collective self-defense. As I have written earlier in the Open Forum, there remains a significant gap in understanding the nature and intent of this change—especially with regard to the broader Japan-ROK relationship. There remains a lack of clarity on the Korean side as to the limitations on the change as well as the legislative steps that are necessary before the reinterpretation can be formally approved. Seoul is concerned about the development of offensive strike capabilities by Japan's Self Defense Forces and seeks clarity on potential exercises of collective self-defense that would involve the Korean Peninsula.

Concerns on China and North Korea

While there was only one session devoted specifically to the topic of Japan-ROK bilateral relations, the issue was an underlying factor throughout, recently marked by concern in Tokyo—and also in Washington—on the ROK's engagement with China. Warming relations between Seoul and Beijing are being closely followed by many who wonder whether President Park Geun-hye has concrete long-term objectives or is simply pursuing a mutually convenient relationship amidst souring ties with Tokyo. Although Park may genuinely aspire to a "strategic cooperative partnership," the notion does not hold up in the medium or long term because Seoul and Beijing have different views on the region's most critical issues, such as the future of the Korean Peninsula and the role of American power in the region. There are also other reasons why Seoul and Beijing are far from budding strategic partners. While both sides have resisted a public dispute, diplomatic tensions remain over China's unilateral declaration of an Air Defense Identification Zone over the East China Sea last November, which includes the Korean-controlled reef Ieodo.

While the Sino-Korean relationship may be limited, it continues to raise concerns not only in Tokyo, but also in Washington, which holds out hope for greater integration of its alliances in the Asia Pacific as part of its rebalance to the region. During his trip to Northeast Asia this past spring, Secretary of Defense Chuck Hagel referred to Japan as the "cornerstone" and the ROK as the "linchpin" of the US rebalance.[1] Despite persistent efforts however, Washington's drive for comprehensive trilateral cooperation with Japan and Korea remains underperforming. While it appears to be gradually accepting the divide between Japan and the ROK, it is not ready to welcome a strategic embrace between China and Korea. The strained trilateral relationship has been a long-held goal for China, which aims to erode the US alliance structure in East Asia—a system that partially is aimed at protecting against Chinese regional assertiveness. This has also flustered attempts to reassure Japan and Korea on US extended deterrence commitments; the two have different threat perceptions—some even at odds with each other—as in the case of Seoul's unease with Japan's defense reforms. These themes were discussed.

Another area that concerns the United States is discussions about North Korea. Generally, Washington welcomes any approach that might get China to alter its firm policy of protecting Pyongyang, but the Obama administration has recognized that it too fell into Beijing's honey pot trap of promising a tougher line against the North. As former US diplomat Joel Wit indicated, "The US should avoid the mistake made after the US-Chinese Sunnyland's summit when many in the US thought that Chinese policy had shifted and Beijing was going to be more proactive in pressuring North Korea to return to the Six-Party Talks and denuclearize."[2]

Is the budding China-Korea relationship likely to have a profound impact on the region? The conference consensus is that we are far away from a strategic partnership that would alter the current regional order. Recent polling from the Asan Institute for Policy Studies in Korea indicates that, while China's popularity is at an all-time high, there remains deep skepticism across the country about China, for example, nearly a third of Koreans still view China as an economic competitor. More telling is the fact that nearly two-thirds of South Koreans see China as a military threat. Similarly, on North Korea, free trade, and support for international institutions, Seoul remains on a different track than Beijing, many concluded.

There was also discussion of the Abe administration's recent engagement with North Korea on the unresolved issue of abducted Japanese nationals. This has been met with caution in Seoul—and Washington—due to fears that it can disrupt trilateral deterrence efforts. In August, Japan's foreign minister Kishida Fumio greeted his counterpart from North Korea, Ri Su-yong, in Myanmar on the sidelines of the ARF—the highest level exchange during the Abe administration. Abe has personified the need to resolve ties with Pyongyang by pointing to the elderly relatives of Japan's still missing citizens. Since retaking office, Abe has repeatedly met with the family members of the abductees and has pressed for resolving the issue with more force and political capital than his predecessors.

Abe's personal investment has motivated him to take more pronounced risks with the aim of resolving the long running saga. During a speech at the Center for Strategic and International Studies in February 2013, Abe addressed the importance of the issue: "Now, if you look at the lapel of my jacket, I put on a blue-ribbon pin. It is to remind myself each and every day that I must bring back the Japanese people who North Korea abducted in the 1970s and '80s."[3] The two sides followed up earlier meetings with working-level talks in Mongolia, including a meeting between the Yokota family (parents of one of the abductees—Megumi Yokota) and their North Korean granddaughter.

These talks culminated in Abe's announcement in July 2014 that Japan would lift some sanctions against the North, focusing narrowly on travel between the countries and the transfer of funds between Japan and North Korea. The Abe government also loosened regulations against North Korean ships entering Japan's ports and promoted the exchange of humanitarian aid. The move was done after Pyongyang agreed to create a Special Commission to look into the unresolved cases of abducted nationals, with a promise to provide Tokyo with detailed reports on their fate. While Abe remains focused on attaining closure, there remain a host of immovable obstacles that will prevent a more comprehensive rapprochement between Tokyo and Pyongyang, conference participants generally agreed. The North's expansion of its ballistic missile program remains one of the chief security concerns for Japan. Despite Kim Jong-un's emphasis on economic reforms, Japan continues to view a "military-first" approach as threatening its security. Moreover, Pyongyang's nuclear and other WMD programs—as well as its significant use of Special Forces, cyber-attacks and espionage—have only strengthened Japan's desire to ratchet up traditional deterrence strategies with South Korea and the United States.

Throughout the past decade, Japan has tried to couple and then decouple these separate issues—and both approaches have their inherent problems. The bundling approach was most clearly defined by Japan's inclusion of abduction negotiations in the Six-Party Talks that concerned the North's nuclear weapons program, drawing the ire of others in the process, who mainly viewed the abduction issue as separate and distracting from the central purpose of the negotiations. With the framework effectively dead, Abe has taken a more aggressive two-track approach. On the one hand, he remains hawkish on Pyongyang's ballistic missile and nuclear programs, which aligns him very closely with Washington and remains the only high-level area of security cooperation with Seoul. On the other, Abe has determined that it is okay to break ranks with the United States and South Korea as long as that path is narrowly focused on the abduction issue.

Looking for Niche Opportunities for Cooperation

Aside from providing a forum for discussion on these concerns, the conference looked at ways in which the Japan-ROK relationship could be improved. By focusing narrowly on historical grievances and bare-minimum cooperation on North Korea, both Tokyo and Seoul are missing out on a host of useful areas of collaboration—many of which already have strong roots but lack the proper nurturing. Promising avenues of nonsensitive cooperation include counterpiracy, energy security, ties between interparliamentary groups, participants noted. These areas can complement the essential drive for unity and information sharing on deterring provocations from Pyongyang. The Japan-Korea Working Group at the Pacific Forum CSIS has been working on these issues over the past year and has looked at ways to transcend the current focus on the political chasm. Of course, a balance can also include the essential cooperation tools on North Korea such as key bilateral agreements: General Security of Military Information Agreement and the Acquisition of Cross-Servicing Agreement. There are other possibilities on missile defense and more joint-operations training to enhance the current level of engagement.

For low-hanging fruit that can build confidence in the relationship, one prime area for cooperation is energy security. According to the World Bank, Japan and the ROK both receive over 80 percent of their primary energy consumption from abroad,[4] making it critical for them to secure a stable supply of energy and resources, to increase their usage efficiencies, and to prevent environment pollution after use. Both also have expertise in strengthening nuclear safety and the development of renewable energy. Magnifying these issues is the global shortage of energy sources, which makes it increasingly important for both to turn to alternative energy sources, such as renewable energy. A second area for cooperation is joint-work on counterpiracy and securing sea-lanes. Japan and Korea are two of the world's most significant trading nations, and each has more than one trillion dollars in global trade annually, most conducted via the seas, making it beneficial to enhance bilateral efforts to guarantee the safety of sea lines of communication. One opportunity is to follow in the footsteps of the US-Canada Shiprider Program: officers from one Coast Guard could be delegated to the vessels of the other for a period of time. Initially, this could be limited to the duration of bilateral or multilateral drills. Later, the number of crewmembers on such "exchanges" and the duration of their stay could increase, and they could become actively involved in the regular operational activities of the other force.

These initiatives need to be complemented by a parallel track at the political level, which would bring Seoul and Tokyo together in a mutually acceptable compromise on their quarrel over historical and territorial issues. In the coming months, it will be vital for both sides to recognize that incremental change is better than no change. A "grand bargain" on all issues may not be realistic now, but both sides can work toward this goal through a reduction of the current trust deficit. In this sense, Japan and Korea should continue to look at nonsensitive areas for enhanced cooperation without focusing on regional tensions. That was the main message from the three-way discussions in Hawaii.

Notes

1. US Department of Defense, "Hagel Discusses US-Japan Security Relations," April 5, 2014, http://www.defense.gov/news/newsarticle.aspx?id=121990.
2. "Summit Reveals Fundamental Differences on North Korea," *Korea Times*, July 4, 2014, http://www.koreatimes.co.kr/www/news/nation/2014/07/485_160385.html.
3. Abe Shinzo, "Japan Is Back: Policy Speech," *Kantei*, February 2013, http://japan.kantei.go.jp/96_abe/statement/201302/22speech_e.html.
4. Shimizu Aiko and Young-June Chung, "Strengthening Energy Cooperation between Japan and Korea," *Issues and Insights, Pacific Forum CSIS* 16, no. 10, 22.

PART II

Japan-ROK Relations under Stress

CHAPTER 5

ROK and US Views on the Foreign Policy of the Abe Administration

Bong Youngshik

Who is Abe Shinzo? Is he a realist who happens to be an historical revisionist, or is he a revisionist whose realism carries the seeds of revived Japanese militarism? South Korean opinion leans toward the latter conclusion, while Americans largely hold the former. Divergent views on the Abe administration and its foreign policy orientations have emerged as a source of friction between Seoul and Washington as security allies.

After Abe clenched a decisive majority in the Upper House elections on July 21, 2013, giving his LDP coalition control over both houses for the first time in decades, United States and ROK analyses regarding Japan's emergent foreign policy have sharply diverged. To many Japan watchers in the United States, the electoral victory was eagerly welcomed for strategic as well as other reasons. After all, it sealed the mutual commitment between Tokyo and Washington to elevate their alliance, which had been expressed during Abe's visit to Washington in February 2013. In personal correspondence, Gilbert Rozman observes that Abe represents four inviting qualities to Americans. First, he is a firm supporter of a stronger alliance, a sharp departure from the Hatoyama Yukio line of 2009–2010, and heir to half a century of LDP endorsement of this relationship with a new, more positive twist centered on upgrading collective defense. Second, he is a strategic thinker about changing security threats, whether from North Korea or China, and takes a supportive attitude toward US thinking on regional reorganization, such as military ties in Southeast Asia and Asia-Pacific economic regionalism through the TPP. Third, Abe is perceived to be a bold leader with an unusual opportunity to command a majority in the two houses of the Diet for more than three years while serving as prime minister. Fourth, Abenomics draws at least cautious praise as the shock treatment that offers the first ray of hope in two decades for shaking Japan's troubled economy from its stagnation, which is viewed as critical for Japan fulfilling its role in security. To US analysts, these merits well outweigh the lone demerit of making offensive remarks or gestures that complicate relations with Japan's neighbors, especially South Korea.

In contrast, to most South Koreans, the right-wing LDP's victory, which came less than a month shy of Korea's Independence Day on August 15, portends more trouble for the already strained relationship between South Korea and Japan. Similar to China, South Korea is alert to Abe's family background, given the legacy of his grandfather Kishi Nobusuke, who was a member of the wartime cabinet and originally included as one of the Class-A war criminals. This colors views of Abe's outspoken desire to rewrite Japan's peace constitution and build stronger, less constrained armed forces. Abe is also judged for his controversial remarks on historical issues, which deepen neighboring countries' suspicions that he would even adhere to past statements, let alone squarely face these issues. During his first term as prime minister in 2006–2007, he denied that there exists concrete evidence to prove that "comfort women" were coerced by the Japanese military. In April 2013, he stated in the Diet that "the definition of what constitutes aggression has yet to be established in academia or in the international community. Things that happened between nations will look differently depending on which side you view them from." South Koreans treat these attitudes as more than insensitivity and nostalgia. They are regarded as indicative of security policies reminiscent of past aggression, posing a threat to regional stability.

There is danger that this clashing interpretation of Abe's intentions and impact will negatively affect security in East Asia and even ROK-US bilateral relations. On the one hand, Americans may lose confidence in South Korea's seriousness about strategic concerns, attributing its response to Japan as emotional nationalism without a foundation in realism. On the other, Koreans could decide that the US response to Japan represents a significant gap in values, even a sign of extreme thinking about China or North Korea which is not in line with Seoul's own thinking. With these concerns in mind, we need to look more closely at the ROK-US divide in reflecting on Abe's impact.

Early Expectations for Abe's Japan

Prior to the Upper House elections in July 2013, the predominant prediction was that foreign policy would be pragmatic and measured, regardless of Abe's personal convictions. There were several considerations supporting this view. First, this is Abe's second chance as prime minister. During his first term, he learned the painful lesson that nationalist fantasies would not ensure political support from the public. Moreover, Abe well knows that Japanese voted for his party in July overwhelmingly in the hope of further economic revival, and the economy must remain his top priority. Japan's economic resurgence depends heavily on its trading partners, China and South Korea included. Thus, one expects that Abe would not let economic policymaking be distracted by needless spats over wartime guilt and historical issues with neighboring countries. As Abe acknowledged in his press conference after winning the Upper House elections, "Economics is the source of national power. Without a strong economy, we cannot have diplomatic influence or strong social security."[1]

Second, Abe's vision of a "strong Japan" has an international dimension. Accomplishing this goal would require more than just arousing inward-looking nationalist sentiment and pride, which, if done in the manner envisioned, would worsen Japan's isolation in the Asian region. Abe has repeatedly offered assurances that he understands how important China is for Japan. After showing that he could moderate his stridency in the interest of easing tensions with China in his first term, he reiterated the same message during his 2012 Lower House election campaign, saying, "China is an indispensable country for the Japanese economy to keep growing. We need to use some wisdom so that political problems will not develop and affect economic issues."[2]

In addition, a main focus of Abe's foreign policy is boosting the Japan-US security alliance. Since the standoff over the Senkaku (Diaoyu) islands escalated in 2012, the United States has been sending a clear message to Japan: it wants "cooler heads to prevail" and opposes any unilateral actions that would make the situation worse. Nationalist actions by the Abe administration in an attempt to bolster its control over the islands would turn Washington away.

Third, Abe's chief foreign policy advisors are pragmatic, notably Yachi Shotaro who expressed reservations about stationing Japanese government workers or Coast Guard personnel on the Senkaku islands to solidify Japan's position on the ground. Such moves, Yachi well understands, would not only be unlikely to deter Chinese intrusions in the adjacent waters, they would likely risk further provocations by China. Yachi also suggested that the Abe administration would take a cautious approach to Dokdo (Takeshima), trying to deescalate tensions created after Lee Myung-bak's visit there in 2012, saying that a decision on whether Japan would try to unilaterally take the Dokdo case to the International Court of Justice would be made after the Abe government was more settled.

Abe's Foreign Policy in 2013: Embracing Washington, Waiting for Seoul

In early 2013, shortly after returning as prime minister, Abe made foreign policy initiatives to both the Obama administration and the Park Geun-hye government. The initiative toward Obama, which was punctuated by Abe's official visit in February, enhanced his image as more of a realist than an historical revisionist, in large part because of his seriousness about boosting Japan's military role. But his initiative toward Park failed to reduce South Korean concern precisely about the revival of Japanese militarism, seen as reassertion of national identity, rather than as realism. The February 22 summit in Washington confirmed Abe's commitment to upgrading the US-Japan security alliance, which he charged had suffered for three years when the DPJ was in power. Abe declared "the trust and the bond in our alliance are back." Obama affirmed that the alliance is "the central foundation for our regional security and so much of what we do in the Pacific region."

Abe made a bold political decision to start the process of Japan joining the TPP, which was at the top of Obama's agenda, despite opposition at home to concessions likely to be required on agricultural protectionism and other sensitive items. He assured the US government that Japan would take action to move the Futenma US military base in Okinawa as promised, and he pledged to increase the national defense budget, a longtime US goal. On the Senkaku issue, Abe assuaged Washington's worry about moves that could increase the likelihood of armed conflict with China, drawing US forces into the fray under the bilateral security treaty, assuring it that he would deal with this issue "in a calm manner" and that he would seek to improve relations with Beijing.

Abe also sought an opportunity to break the stalemate between Seoul and Tokyo. Optimists claimed that past strains between Lee Myung-bak and Noda Yoshihiko were caused by an uncertain political structure and lame duck leadership. Therefore, both countries now had an opportunity for a fresh start under newly elected leaders. In January 2013, Nukaga Fukushiro, a lawmaker with long ties to South Korea, met president-elect Park as a special envoy and delivered a letter, in which Abe promised to work closely with the ROK government and to deepen the bilateral relationship from the standpoint of a "big country." He mentioned the "past success" of bilateral relations to which his grandfather and her father had contributed, based upon deep personal trust between the leaders of Korea and Japan. Abe also praised South Korea as "a very important neighboring country that shares the values of democracy and a market economy," while calling for expanded security ties with other democracies in the region, including India and Australia, to offset China's growing presence.

Park's response agreed with Abe that trust is essential in promoting partnership between the two countries, but she made it clear that South Korea's cooperation would depend on Abe's policies on sensitive historical and territorial issues. For instance, in response to the special envoy's question on what was her secret in becoming known as the "queen of elections," Park made a reference to the Confucius saying that no leader can stand without the trust of the people. She added that public sentiment in South Korea on historical issues should be respected in building a forward-looking partnership.

Cautious hope for an improvement in the bilateral relationship was quickly dashed. In April, members of the Abe government paid visits to the Yasukuni Shrine. In protest, the ROK Foreign Ministry cancelled the scheduled ministerial meeting in Tokyo. The first meeting between foreign ministers of both countries had to wait until July 2013, when the ROK foreign minister Yun Byung-se and his Japanese counterpart Kishida Fumio met at the ASEAN Regional Forum in Brunei. As for a summit meeting between Park and Abe, it remains only a remote possibility.

Japan and South Korea in the State of Mutual Abandonment

Although many observers point to the idiosyncrasy of top leaders and their personal views of history and politics, the prolonged stalemate, even deterioration, in the ROK-Japan relationship over the past decade suggests that there is a structural force at work. One can go so far as to conclude that the two no longer find each other to be useful partners. Instead, they increasingly view the other as a nonfactor or even impediment in pursuing their respective national interests. Apart from the start-up period of DPJ rule, relations have been almost continuously troubled since Koizumi Junichiro became prime minister and visited the Yasukuni Shrine in 2001. Despite criticism from both the domestic and international community, he continued to visit the shrine—a total of six times during his five years and five months holding that post.

A ray of hope emerged when Hatoyama Yukio took office in 2009 with a foreign policy vision to create a more harmonious Asian community. However, he resigned less than a year later, and, subsequently, Kan Naoto and Noda failed to find common ground with Lee Myung-bak. In fact, South Korea and Japan faced a diplomatic crisis in 2012 amid renewed tension over the portrayal of history and territorial claims over Dokdo (Takeshima). To Japan's dismay, Lee became the first South Korean leader to visit the island in August. Afterward, Lee commented that if the emperor of Japan wanted to visit South Korea, he should first sincerely apologize for Japan's past colonial rule, sparking more hostility between the two countries. There has been no ray of light over more than a year, even after leadership changed in both countries.

Why are there no signs that the Abe administration or the Park government will be proactive in trying to overcome the prolonged stalemate? Arguably, neither South Korea nor Japan is in desperate need of each other for pursuing its essential national interests. They lack the incentive to energetically try to improve bilateral relations. One could draw a parallel with the relationship between Roh Moo-hyun and George W. Bush in 2003, a year marking half a century of alliance, but in reality, one of the most turbulent times in the relationship. Intensive anti-US sentiments had flared in Korea after two schoolgirls were accidently killed by a US army vehicle. The soldiers responsible for the incident were acquitted by a US military trial, sparking a massive candlelight rally amid public sentiment that it was time for Korea to become truly independent from foreign influence. In addition, the possibility of a preemptive strike against North Korea, which had expelled the IAEA inspectors and was restarting its nuclear reactor, raised fear among South Koreans that their country would become the victim of North Korean retaliation. In this light, the alliance was seen not as an asset for South Korea but as a threat to its survival. Thus, it was believed by many Koreans that South Korea would need to distance itself, or even abandon the United States all together, to achieve national stability and reunification.

The United States, in contrast, expressed concern that the Korean government did not take a stronger stance to stop the anti-US demonstrations. Washington even questioned whether South Korea could be trusted as an ally. Despite the military standoff with North Korea and clear evidence that it was developing nuclear weapons, Seoul kept its distance from Washington and approached Pyongyang with the aim of reunification. Some Americans suggested "abandoning Korea" amid resentment that it was no longer grateful for the sacrifices the United States had made to keep peace on the peninsula.[3] Relations between the two countries only started to recover after 2004, mainly because both concluded that they needed cooperation to enhance their own national interests in a context that had shifted to the Six-Party Talks.

The current ROK-Japan relations are similar to those ROK-US relations in their state of "mutual abandonment." During the Cold War, South Korea and Japan needed each other for security against common adversaries and economic prosperity. They shared the goal of preventing the spread of communism in Northeast Asia and an outbreak of war. Yet in this post–Cold War era, they have found it difficult to determine which security threat should be their common priority and how they should modify their policies accordingly. Seoul and Tokyo have different strategic calculations, the former in addressing North Korea's nuclear threat and preventing its aggression. If the South Korean government believed that cooperation with Japan were essential to deter North Korea, it would have become more cooperative long ago. Instead, it views strengthening security ties with the United States and improving relations with China as sufficient, in part because they actually can exert influence over North Korea. This seems more promising than pursuing ties with Japan, with whom the effectiveness of security cooperation has yet to be proven. The same could be said about economic cooperation. Japan continues to be an important trade partner, but it is no longer as important as it was back in the 1997 financial crisis.

The End of Japan's "Deference Diplomacy"

As both countries entered a state of "mutual abandonment," political parties and civil society on each side lost interest in taking initiatives to improve bilateral ties. On the surface, the two administrations claim that they want to improve bilateral relations, but, in reality, their attitudes convey the message that they simply want to minimize any outbursts, given that political cooperation is not a top priority at this moment. Seoul is engaging in "wait-and-see diplomacy," asking Japan first to apologize for its historical conduct and show sincere willingness to resolve the situation.[4] Japan claims that it is making an effort to address issues regarding its portrayal of history and Dokdo (Takeshima), but political and intellectual leaders have lost enthusiasm for "deference diplomacy," believing that Seoul is unwilling to change its stance no matter what kind of goodwill gesture and incentives Japan might offer.

There is a kind of apology fatigue. Believing that Japan has long tried "deference diplomacy" toward Seoul, which has ended with disappointing result, many consider it futile to try again to meet Seoul's changing expectations.

In the eyes of Koreans, the Japanese government has long claimed that it wanted to improve relations, but it always succumbed to the pressure of domestic politics to pander to the conservative base. As a result, overtures toward Korea have tended to be half-hearted and insincere, falling short of Korea's expectations. For instance, one of Abe's campaign promises in the December 2012 general elections was to elevate "Takeshima Day" into a national event supported by the central government. Wary that this would provoke South Korea, the Abe administration tried plausible deniability. It did not send any high-ranking cabinet member to the "Takeshima Day" event in Shimane prefecture. Instead, it sent the Cabinet Office's Parliament Secretary Shimajiri Aiko. From South Korea's standpoint, this was tantamount to the central government supporting the ceremony as a national event, since it was the first time a senior government official had been sent to attend the ceremony. But from the standpoint of the Abe government, the decision should be viewed as evidence that the government took the South Korean demands under most serious consideration for the sake of improving bilateral ties.

In August 2013, China and South Korea sent a clear message to the Abe government that visits to the Yasukuni Shrine by high-ranking cabinet members, including Abe himself, would be regarded as the government's refusal to view Japan's past history objectively. Abe left the issue to individual cabinet members, by announcing that he would not "force" any to decide one way or another on visiting the shrine on August 15.[5] Abe and three core officials—Deputy Prime Minister Aso Taro, Chief Cabinet Secretary Suga Yoshihide, and Foreign Minister Kishida Fumio—did not visit it, which in Abe's mind should be considered as sufficient to take into account concerns expressed by Seoul and Beijing. But he also sent an offering instead, which made the Asian countries question the wholeness and sincerity of his decision. He also broke away from tradition in marking the sixty-eighth anniversary of Japan's surrender, by neglecting to explicitly mention its wartime aggression, and he skipped a formulaic pledge that Japan would never again go to war. At the start of an annual autumn festival in October 2013, Abe once again refused the call from the conservative base to visit the Yasukuni Shrine, but the impact of such a conciliatory gesture was quickly cancelled out as he expressed his "regret" for not visiting the shrine and more than 160 governmental officials and Diet members paid a visit. Such calibrated attempts to appease both domestic and foreign audiences raised more serious questions about the sincerity of Abe's proclaimed goal of reviving strained South Korean ties.

Perceptions of Japan abandoning its traditional "deference diplomacy" negatively affect South Korean views of Japan. In fact, a survey by the Asan Institute for Policy Studies in July 2013 indicated very negative impressions of Japan and the Abe administration. People were asked to rank nations according to four factors: attractiveness, trustworthiness, political influence, and the country's leader (see appendix). Results on attractiveness indicated that Japan fell to an average of three out of ten possible points over the past year. In contrast, China reached an average of almost five points immediately following Park's visit, 24 percent higher than the prior year (see table 5.1).

In terms of trustworthiness, the questionnaire revealed that Koreans trusted China (31.7 percent) more than Japan (11.4 percent) (see table 5.2). Among foreign leaders, Koreans ranked Obama first (6.29 of 10), then Xi Jinping (5.35), followed by Putin (4.08). Abe's 1.65 rating was the lowest next to Kim Jong-un (1.14) (see table 5.3). As the negative impressions of neighboring countries each year have barely changed, it is unlikely that icy Japan relations will thaw anytime soon.

The Japan Factor in the ROK-US Alliance

While the Obama administration has embraced Abe as a key partner in its rebalancing strategy toward Asia, Seoul has sought Washington's understanding about its concerns regarding the expansion of Japan's defense forces in regional security affairs without due acknowledgment of its past wrongdoings and militarism. But it was not successful, and Japan's enlarged role in regional security under the principle of collective self-defense appears to be a fait accompli. The 2 + 2 meeting in Tokyo in October 2013 produced an agreement to broaden the scope of the security alliance by expanding Japan's role, positioning surveillance drones in Japan for the first time, launching cybersecurity projects, and placing a new X-band radar system in Kyogamisaki.

It appears that the Park government has shifted its diplomatic focus from getting Washington's agreement that the Abe government's reassessment of the colonial period be a precondition for Japan to push the bounds of constitutional constraints on its military, including the right to collective self-defense, to continually conveying concern to Washington that Abe not attempt "*tongmi bongnam*" (use the United

States to contain South Korea) in order to dissuade Seoul from its basic principles of diplomacy.

In the past, perceptions that the United States is taking Japan's side have triggered anti-American sentiment in Korea. In February 2005, when Japan's Shimane prefecture stirred controversy by passing a bill designating "Takeshima Day," South Korea's National Security Council reacted strongly by announcing a "New Doctrine on Dealings with Japan." Roh Moo-hyun's expression of this concern affected South Korea-US relations as well. When in March then US secretary of state Condoleezza Rice was in Seoul to coordinate responses to North Korea's nuclear weapons program, Roh devoted 45 minutes of an 80-minute meeting to discussing Korea's problems with Japan. In June 2006, Roh also used his summit meeting with George W. Bush to criticize Japan's handling of its past war crimes and colonial rule, including inaccurate portrayal of history in its textbooks.[6]

As seen in Seoul, unless it is kept in harmony with efforts to upgrade the ROK-US security alliance, the Obama administration's effort to increase the contribution of Japan's SDF to regional security in Northeast Asia can strain ROK-US relations. No matter how confident US officials may feel that vigilance accompanied by quiet diplomacy will be sufficient to dissuade the Abe government from provocative moves regarding history toward China and South Korea, South Korea's mistrust in Japan may be deeper and more resilient than they assume. During her meeting on September 30, 2013, with Defense Secretary Chuck Hagel, who expressed hope for improved ties between South Korea and Japan, Park criticized the Japanese leadership repeatedly for its regressive remarks on history and territorial issues. She added that security cooperation between South Korea and Japan would be limited unless Japan earns the trust of South Korea.

Given deep-seated reservations about Japan playing a larger role in regional security affairs, it is unlikely that South Korea will be agreeable to forging an ROK-US-Japan virtual alliance in the near future, despite US desire to synergize its alliances with the two key pillars of its rebalancing policy toward Asia. Strong resistance and distrust in the ROK leadership and general public toward Japan are likely to remain a major obstacle to a central part of US security strategy.

Conclusion

The US rebalancing policy in Asia has had an opposite effect on Japan and South Korea. For Abe's Japan, US policy created a welcome opportunity to loosen the constitutional constraints on the SDF and significantly increase its defense posture. For Park's government, there is concern that the strategic autonomy of South Korea is being diminished in the process of strengthening the US-Japan alliance. It was reported in the Korean media that a deal was struck during Hagel's visit to Seoul that, in exchange for US acceptance of the request to postpone the return of wartime operational control to South Korea, it had agreed to join the US-led missile defense system. If South Korea will introduce the Terminal High Altitude Area Defense system (THAAD), a land-based system designed to destroy incoming missiles during their final phase of flight, and increase the interoperability of the Korea Air and Missile Defense (KAMD) system, China will certainly view this as joining in its encirclement by the United States and Japan.[7]

It has been reported that the Xi Jinping government, which initially had high hopes after the Beijing summit in June that the Park administration would become a closer partner than its predecessor, grew disappointed as it viewed South Korea moving closer to accepting US demands for participation in missile defense and the TPP.

South Korea has long evaluated its security partnership with the United States in two ways. On the one hand, South Koreans assess it in terms of its capacity to meet threats to security interests. On the other, they assess the partnership in terms of fairness, normally through comparisons with the US-Japan alliance. This means that the US position on historical and territorial issues left from the colonial period has an important bearing on South Korea's perception of the alliance. The "pivot" to Asia has found a supportive partner in Abe, but the broadening of the US-Japan alliance leaves doubt in South Korea that US leaders may prioritize interests in the region that do not correspond to South Korea's priorities. The divergence over Japan in 2013 is proving to be a test of the ROK-US alliance, which may be a harbinger of further tests in the coming years.

The Park administration has already expressed its displeasure and sought assurances from the United States that if the US-Japan defense cooperation guidelines are revised in ways that recognize Japan's right to collective self-defense, South Korea's sovereignty would not be affected. This reflects a deep-seated concern that, in case US forces in South Korea come under attack from North Korea, Japan could exercise its right of self-defense and send its Self-Defense Forces into South Korean territory.[8]

US readers may ask: If the objective of the US-Japan alliance is to deter threats from a nuclear-armed North Korea and potential threats from China in the midst of an unprecedented arms buildup in the region, why do South Koreans not take those threats seriously? Are South Koreans allowing an obsession with an historical identity issue to color their judgment and overshadow what many regard as a realist response to the emerging security environment? The answer South Koreans are giving is this: Historical experience convinces them that Japan's military resurgence, even starting on a scale that does not raise eyebrows in most circles, poses a realist concern. This is the message that in the fall of 2013 is rattling ROK-US relations.

Appendix

Table 5.1 A nation's attractiveness

Question: With 0 for "not attractive at all," 5 for "average," and 10 for "very attractive," how favorable do you regard [US/China/Japan/North Korea]? Pick a number between 0 and 10. [Unit: Average, 11-Point Scale]

Survey period	August 14–16, 2012	January 3–5, 2013	March 30– April 1, 2013	May 2–4, 2013	July 1–3, 2013
US	5.55	5.70	5.81	5.74	5.61
Japan	2.93	3.31	3.19	2.93	2.96
China	3.97	4.45	4.38	4.22	4.92
North Korea	3.19	2.99	2.03	2.07	2.27

Table 5.2 Country trustworthiness

Question: How do you rank each country's trustworthiness? (Survey Period: July 7–9, 2013) [Unit: %]

	Trustworthy	Don't trust	Don't know/no answer
US	57.3	37.0	5.70
European Union	48.0	31.2	20.8
China	31.7	61.8	6.50
Russia	19.9	58.7	21.4
Israel	19.0	50.9	30.1
Japan	11.4	85.0	3.60
Iran	6.10	70.9	23.0
North Korea	5.60	90.2	4.20

Table 5.3 Favorability toward a nation's leader

Question: How do you view each nation's leader? (Survey Period: July 10–12, 2013). Please rank your responses according to: 0–4 points, not favorable; 5 points, average; 6–10 points: favorable.

	Favorable (%)	Average (%)	Not favorable (%)	Don't know/no response (%)	Total average (points)
US President Obama	56.9	23.8	11.6	7.70	6.29
China's General Secretary Xi Jinping	35.4	24.1	21.7	18.8	5.35
Russia's President Putin	14.0	25.0	33.4	27.6	4.08
Japan's Prime Minister Abe	5.00	9.00	76.5	9.50	1.65
North Korea's Kim Jong-un	2.30	6.30	85.1	6.20	1.14

Source: The Asan Institute for Policy Studies, Asan Poll, July 1–15, 2013. The sample size of each survey was 1,000 respondents over the age of 19. The surveys were conducted by Research & Research, and the margin of error is +/-3.1% at the 95% confidence level. All surveys employed the Random Digit Dialing method for mobile and landline telephones.

Notes

1. Jonathan Soble, "Abe Issues Party Rallying Cry after Poll Win," *Financial Times*, July 22, 2013.
2. Mark McDonald, "To Japan-China Row, Add One Potential Provocateur," *International Herald Tribune*, December 19, 2012.
3. Nicholas Eberstadt, "Our Other Korea Problem," *The National Interest* (Fall 2002); Richard V. Allen, "Seoul's Choice: The U.S. or the North," *The New York Times*, January 16, 2003.
4. "Editorial: Problem Is Prime Minister Abe's Honestly," *Choongang Daily*, July 30, 2013.
5. "Abe Gives Government Officials Freedom to Visit Shrine on 8.15," *Donga Ilbo*, August 7, 2013.
6. Yoichi Funabashi, *The Peninsula Question: A Chronicle of the Second Korean Nuclear Crisis* (Washington, DC: Brookings Institution Press, 2008), 261.
7. "ROK-US Made Progress on MD-OPLAN Control Big Deal," *Moonhwa Ilbo*, October 15, 2013.
8. "S. Korea Seeks Input in US-Japan Defense Cooperation Guidelines," *The Hangyorye*, October 28, 2013.

CHAPTER 6

Japan's Defense Reforms and Korean Perceptions of Japan's Collective Self-Defense

J. Berkshire Miller

The past few months (to February 2014) have seen a wide scope of long-anticipated defense and security reforms take shape in Japan. Specifically, last December, the administration of Abe Shinzo released its first-ever National Security Strategy (NSS) and announced the creation of a newly minted National Security Council (NSC). Additionally, Japan released newly revised National Defense Program Guidelines (NDPG) on December 17. While there was some vociferous opposition in Beijing to Tokyo's national security reforms, criticism from Seoul has been considerably more nuanced and subtle. However, the most controversial of these reforms, and one that has not yet been enacted, is Abe's push to reinterpret Japan's right to collective self-defense.

The difference between constitutional revision and the reinterpretation of collective self-defense is dramatic but not well understood. Indeed, South Korean diplomats reportedly warned Japan to tread lightly on this reinterpretation and cautioned it to hesitate in taking actions against North Korea under this new authority. Some have even gone further and claimed that they would view such an attack on the North, which the South recognizes as its sovereign territory, as an act of war with Seoul.[1] This is a sentiment that baffles most in Japan as indicated by a senior defense official who, at an Asan conference on "US-Korea-Japan Trilateral Extended Deterrence" this past fall, remarked that "Japan has never even contemplated sending troops to Korea (in the event of hostilities)."

Seoul's fears about Japan's defense reforms and collective self-defense also run contrary to the general US, along with other key partners such as the United Kingdom and Australia, support for these changes. Perhaps even more concerning, however, is the potential damage that these frayed ties with Japan may inflict upon Washington's desire to maintain healthy alliances with both Tokyo and Seoul. Indeed, while Japan remains the cornerstone—in political and security terms—of the "rebalance," it remains essential for Washington to also nurture its long-standing alliance with South Korea. It would be dangerous to dismiss the implications of this rift on Washington's regional goals. As Bong Youngshik wrote in Chapter 5, "There is danger that this clashing interpretation of Abe's intentions and impact will negatively affect security in East Asia and even ROK-US bilateral relations. On the one hand, Americans may lose confidence in South Korea's seriousness about strategic concerns, attributing its response to Japan as emotional nationalism without a foundation in realism. On the other, Koreans could decide that the US response to Japan represents a significant gap in values, even a sign of extreme thinking about China or North Korea which is not in line with Seoul's own thinking."[2]

While Japan and the United States may disagree with Korea's interpretation of Abe's intentions for reforming Japan's defense posture, it would be careless to disregard these concerns. A deeper understanding of Seoul's fears is essential not only to maintain the health of both key alliances, but also to enhance much needed trilateral cooperation on common threats such as North Korea. This chapter analyzes Japan's reforms along with their interpretation by Korea. The current mainstream discourse suggests that Korean perceptions are merely a reflection of the current geopolitical divide between the Park and Abe administrations. In reality, this is only a partial truth. There are several other significant factors, such as the recent amelioration of Korea-China ties that need to be examined.

Japan's Defense Reforms

Before analyzing the concerns from Seoul on Japan's security and defense changes, it is important first to briefly outline the main changes under way in Tokyo. The proposal on collective self-defense is complemented by a host of other recent security and defense reforms, including release of the NSS, a new bill on protecting against leaks of classified information, and the revised NDPG. These changes buttressed a new streamlined NSC in Japan, which is now operational. While there are kinks to be worked out with the NSC, it remains essential for instant and critical national security decisions affecting Japan's national security. The prior arrangement was not as centralized and led to several streams of information arriving at the prime minister's desk through separate channels. Often contradictory and opaque, this resulted in chaotic and unclear decision-making in the cabinet. Now there is one advisor over the entire process, the veteran foreign ministry official Yachi Shotaro, with whom Abe has long-standing relations.

The new defense reforms will bolster the SDF's ability not only to respond to regional and national security issues, such as those in the East China Sea and the Korean Peninsula, but also to assist emerging allies in other regions of Asia. A clear example is the current humanitarian assistance and disaster relief efforts the SDF is undertaking in the wake of typhoon Haiyan in the Philippines. This work builds trust in the new role of the SDF in other regions of Asia and should demonstrate the positive role that the SDF can play in international peace and security.

One of the most controversial, but least likely, areas of reform is constitutional revision. For example, amending Article 9 of the Japanese Constitution, which renounces Japan's right to war as a sovereign nation, requires a two-thirds vote in both houses of the Diet and a majority vote in a public referendum. Despite controlling both houses, "the two-thirds bar remains outside of Abe's reach and his party relies on at least a modicum of compromise and deference from the coalition partner, the New Komeito Party, which holds enough seats to prevent the LDP from getting a two-thirds majority in the Lower House if it withdraws support for the prime minister."[3]

While constitutional revision remains a long shot, Abe is pressing forward on constitutional reinterpretation in order to broaden the mandate of the SDF, assembling an Advisory Panel on Collective Self-Defense to draft a report on whether the government should reinterpret Japan's right to engage in collective self-defense. This is a right that all states maintain through the United Nations charter, but Japan has a self-moratorium since the end of World War II. The panel has already recommended changing the interpretation of the constitution in order to allow the SDF to expand its role abroad in such missions. The panel has also looked to quiet concerns from Japan's neighbors—especially China and South Korea—that this will result in a rebirth of Japan's militaristic past. In this respect, the draft report limits the scope of collective self-defense to include: armed attacks launched on Japan; situations when there are no other appropriate means to respond; and any actions taken in exercising the minimum necessary use of the right.[4]

Of course, the absence of collective self-defense has also been a point of contention in the alliance with the United States as it effectively prohibits Tokyo from responding to a potential third party attack on its ally. Under the current interpretation, if North Korea were to fire a ballistic missile over Japanese airspace toward the US mainland, Tokyo would be unable to assist its ally. Washington's long-term plans to hedge against China's rise with a more robust alliance partner in Japan are also at risk. As Evan S. Medeiros, senior director for Asian affairs at the National Security Council, once remarked, "The US-Japanese alliance is the most important and long-standing element of US security strategy in Asia and is central to efforts to hedge against the possible emergence of a revisionist China."[5]

Yasukuni Clouds the Skies

A critical error in recent analysis on Japan's security reforms is the perception that this is "all about Abe." Admittedly, Abe has helped to fuel this view through his insistence on visiting the Yasukuni Shrine in December 2013. It seems that Abe obstinately refused to listen to pleas from his advisors and allies. According to *Yomiuri Shimbun*, Abe dismissed such warnings, telling his aides that "even if I pay a visit to Yasukuni, they (ties with South Korea and China) won't deteriorate further. Japan has established a good relationship with Russia and other countries aside from those two countries."[6] Abe's inability to resist his personal desire to go to Yasukuni will result in a number of political and diplomatic repercussions at home and abroad.

In Japan, Abe dismissed strong initial opposition from his chief cabinet secretary Suga Yoshihide, and left two of his key ministers—Kishida Fumio (Foreign Affairs) and Onodera Itsunori (Defense)—only a few hours to prepare for the diplomatic repercussions. Abe's decision also will likely have political consequences as he spurned criticism from Yamaguchi Natsuo, the leader of the Komeito, and, during a telephone exchange on the morning of his visit, he rebuffed Yamaguchi's call to avoid the shrine, reportedly saying, "I'll visit the shrine at my own discretion."[7] This could result in blowback if the administration tries to push through more security and defense reforms, especially those requiring legislative approval. The Komeito has already been deeply critical of Abe's plans for constitutional reinterpretation or revision.

The Yasukuni visit has also resulted in significant blowback abroad. China and South Korea will use the trip as more ammunition for their arguments that Tokyo remains recalcitrant on coming to terms with its role in World War II. The Chinese foreign ministry even went as far as decrying Abe as "celebrating the Nazis of Asia" through his visit.[8] Seoul used less bombast, but still levied an accusation at Abe for "digging up wounds of the past."[9] Moreover, according to a report by the *Asahi Shimbun*, Abe's visit derailed the quiet planning that had been going on for months on arranging a summit between Abe and Park.[10] The move may also derail recent efforts to progress on trilateral free trade talks between the three countries. North Korea's state news agency meanwhile, in its predictably over-the-top manner, labeled the decision an "act of war on Asia." Worst of all, Abe's decision has bailed out both China and South Korea, which were being blamed by Washington for their recent policies of aggression and isolation, respectively, toward Tokyo. The visit resulted in an unusual US rebuke, calling the decision "disappointing" and one that could "heighten regional tensions."[11]

The visit to Yasukuni overshadowed some positive developments on security between Tokyo and Seoul over the past year, most importantly, cooperation on North Korea. Last June, there was a trilateral defense ministers meeting arranged by Washington, on the sidelines of the Shangri-La Dialogue, where all sides reaffirmed the importance of "trilateral cooperation, based on common values and shared security interests, and their nations' cooperative efforts to contribute to peace and stability in the Asia-Pacific region and around the globe."[12]

Contextualizing Collective Self-Defense

Seoul's critiques of Japan's evolving defense posture are most acute on the issue of collective self-defense; however concerns over Japan's role in the event of hostilities on the Korean Peninsula are long-standing and not merely an outgrowth of Abe's plans for collective self-defense. In 1998, the Diet passed a law concerning "Measures to Ensure the Peace and Security of Japan in Situations in Areas Surrounding Japan."[13] This legislation essentially allows the SDF to respond to threats in Northeast Asia that could threaten Japan's security. It is largely an operational tool that serves the US-Japan alliance by allowing the SDF to adopt important measures such as rear area support, rear area search and rescue operations, and ship inspection operations.[14] These measures, including the provision of goods, services, transportation, and medical services, are essential support complementing the operations of US forces in situations surrounding Japan, as stipulated in the Japan-US Security Treaty.

In 2007, during his first tenure as prime minister, Abe commissioned an advisory panel to consider whether it made sense for Japan to reinterpret its right to collective self-defense. The Advisory Panel on Reconstruction of the Legal Basis for Security focused its recommendations on four case studies: (1) defense of US naval vessels on the high seas; (2) interception of a ballistic missile targeting the United States; (3) use of force during peacekeeping operations (PKO); and (4) logistical support for PKO.[15] The conclusion was that Japan would endanger its own security, as demonstrated in each of these cases, if it did not reinterpret its right to collective self-defense. The report also questioned the impact that maintaining the ban would have on the future health of the US-Japan alliance as well as on Tokyo's role as an active contributor to international peace and security. Abe's stunted first term, however, prevented him from following through on the recommendations from the advisory panel.

Alliance burden-sharing and the SDF's ability to contribute in joint operations has long been a sticking point in the US-Japan alliance. Hosoya Yuichi recently noted, "Many security experts have seen Japan's self-imposed ban on the right to exercise collective self-defense as a constraint to deepening the US-Japan alliance."[16] Moves to "normalize" Japan's defense posture have been welcomed in Washington as indications that Japan can be counted on not just diplomatically and economically, but also militarily. It is generally highly supportive of Abe's moves on collective self-defense and other security reforms, in line with the ongoing revision of defense guidelines with Tokyo under the alliance. The guidelines, which have not been revised since 1997, are likely to focus on building enhanced cooperation in "areas surrounding Japan" on intelligence, surveillance, and reconnaissance (ISR) as well as on efforts to enhance deterrence.[17]

There is recent divergence on threat perceptions, Japan being primarily anxious about China's actions to change the status quo surrounding the Senkaku/Diaoyu islands, and the United States, while wary of China's increased assertiveness in the East and South China seas, and South Korea would like to see the alliance focus more on the threat emanating from North Korea. While neither would likely support deploying the SDF to the Korean Peninsula in a potential conflict, the situation becomes more complicated with regard to targeting North Korean vessels, rockets, and aircraft outside Korean territory. Under the current interpretation of the 1998 law, the SDF can offer the United States a range of rear area support options but could not come to the aid of US ships and planes that were being attack from outside Japan's territory. A reinterpretation of collective self-defense would enable Tokyo to remove these operational handcuffs.

Implications for Korea

Abe's push to reinterpret collective self-defense presents concerns to Korea for four interlinked reasons. First, Seoul is wary that a reinterpretation could be used as a pretext to use military force against North Korea and potentially even deploy the SDF to the Korean Peninsula in the event of conflict. Second, there is concern that Abe's moves on collective self-defense mask grander strategic goals from Tokyo to significantly build up its military. Third, there is a sense of uncertainty on the effect these proposed changes may have on Korea's alliance with the United States and also its implications for a potential regional arms race. Finally, there is concern that such moves by Japan will widen the divide between the China and US "camps," diminishing Seoul's ability to hedge.

With regard to potential Japanese military action against North Korea, Seoul has also alerted Japan not to extend this reinterpretation to collective self-defense exercises involving the Korean Peninsula. Last November, during a Japan-Korea defense vice ministers' meeting, Korean vice minister Baek Seung-joo cautioned his counterpart on this point.[18] Seoul's second concern relates to a fear that Abe is intentionally using the initiative as a geopolitical tool to reorient Japan toward a more muscular military. This has also raised fears that Abe's intentions are not focused on the US-Japan alliance or participation in PKOs overseas. These concerns lead to the third and fourth set of potential implications, which could reorient the strategic landscape in Asia and undermine both Korea's alliance with the United States and its delicate balancing act with China.

After the 1998 law came into force, the Korean government immediately sought assurances from Tokyo that it would not be used as a pretext for deployment of the SDF against North Korea. It "declared that its permission was necessary for the Japanese Self-Defense Forces to launch a military operation in South Korea's territorial and maritime area (a rear-area support for the US). The South Korean government still demands that Japan must obtain permission, which it considers an appropriate measure."[19] It has been 15 years, but these concerns remain.

The linkage between Abe's views on history and Japan's normalization in military-security terms is perhaps the most problematic issue threatening a long-term rapprochement between Tokyo and Seoul.[20] Park Young-June recently warned of the dangers of subscribing to this erroneous nexus: "Concerning the future direction of South Korea's response, the South Korean government must first objectively and dispassionately understand the purpose behind the change in Japan's national defense policy. The Abe administration is moving to become a normal country—not a militaristic—country. It is not a proper response to overreact against Japan's exercise of the right of collective self-defense under an assumption that it would soon result in an encroachment of South Korea's sovereignty."[21]

Framing Japan under Abe as a potential security threat is an alarming development. Bong Youngshik noted in Chapter 5, "It appears that the Park government has shifted its diplomatic focus from getting Washington's agreement that the Abe government's reassessment of the colonial period be a precondition for Japan to push the bounds of constitutional constraints on its military, including the right to collective self-defense, to continually conveying concern to Washington that Abe not attempt 'tongmi bongnam' (use the United States to contain South Korea) in order to dissuade Seoul from its basic principles of diplomacy."[22] To officials in Tokyo and Washington, this fear has little grounding. First, South Korea's alliance with the United States is an essential pillar of Washington's "rebalance." Second, it seems like an imaginative stretch to see Abe's courting of the US-Japan alliance as a tool to contain Seoul, especially in light of Tokyo's near-obsessive focus on mitigating Chinese provocations in the East China Sea.

Concern over tongmi bongnam, whether logical or not, persists in many policy circles in South Korea. Robert Dujarric recently alluded to this concerning development in Seoul's posture toward Tokyo, "South Korea is a US ally with the same broad geostrategic objectives as Japan. Unfortunately, many South Korean politicians have used Japan as a political football at home. Some even give credence to the hoax that Japan poses a military threat to Dokdo (Takeshima). North Korea is a menace to the South; Japan is not."[23] Sukjoon Yoon also recently stressed that "South Korean national military strategy no longer revolves around ideological and military confrontation with the North, but should focus primarily upon potential maritime conflicts with maritime powers."[24] Yoon proceeded to point to Japan as one example, citing its launch of the Izumo—a massive-helicopter carrying destroyer revealed this past summer.

Conclusion

Korean suspicion of Tokyo's plans on collective self-defense and its renewal of its National Defense Program Guidelines pose an unnecessary irritant to the US-Japan alliance, which remains the security cornerstone of Washington's rebalance. There are some positive signs. Park Young-June recently stressed that "under the framework of the US-Japan alliance, the revival of Japanese militarism is unlikely."[25] Likewise, Kim Heung-kyu called upon Seoul to "refrain from making emotional responses and explore ways to enhance our national strategic interests."[26]

Japan's defense and security reforms are long-standing needs for the US-Japan alliance. Despite this, it remains crucial for the Abe administration to engage in a strategic communications plan that will help dispel the idea that reforming the SDF goes hand in hand with a more threatening posture to Japan's neighbors. Engaging in more bilateral and trilateral (with the United States) dialogues with a specific focus on this issue would allow Japanese to explain their goals while the Koreans can outline their concerns in more detail and without the political risk of public meetings.

A poll last year points out that 62 percent of South Koreans feel threatened militarily by Japan. A more recent poll following Abe's visit to the Yasukuni Shrine indicates that over 85 percent of Koreans feel that the relationship with Japan is "bad." Indeed, polling has even suggested that Abe's image is worse than that of Kim Jong-un. Despite these troublesome indicators, a near majority (49.5 percent) of South Koreans feel there is a need to have a leader's summit with Japan. Before the Yasukuni visit, the number was closer to 60 percent. Similarly, even after Abe's shrine visit, most Koreans agree on the necessity of signing a General Security of Military Information Agreement with Japan.[27] A leaders' summit in 2014, if carefully planned, could diffuse some of the biggest thorns in the relationship. Given these ambivalent polling numbers, most South Koreans would likely react favorably if an upbeat mood were achieved, while Japanese would undoubtedly welcome it.

Notes

1. A senior Korean diplomat told the *JooongAng Daily* that "Japan should not jeopardize the regional security and peace, which holds paramount value [through the exercise of its right of collective self-defense]. And Japan will be held accountable for proving whether their deployment of armed forces are justifiable or not."

2. Bong Youngshik, "ROK and US Views on the Foreign Policy of the Abe Administration," *The Asan Forum*, November 6, 2013, http://www.theasan forum.org/rok-and-us-views-on-the-foreign-policy-of-the-abe-administration/.

3. J. Berkshire Miller and Takashi Yokota, "No About-Face for Abe," *Foreign Affairs*, July 30, 2013, http://www.foreignaffairs.com/articles/139610/j-berkshire -miller-and-takashi-yokota/no-about-face-for-abe.

4. "Minimum Change Eyed to View on Collective Self-Defense Right," *Yomiuri Shimbun*, December 2, 2013, http://the-japan-news.com/news /article/0000838087.

5. Evan S. Medeiros, "Strategic Hedging and the Future of Asia-Pacific Stability," *The Washington Quarterly* 29, no. 1 (Winter 2005–2006): 145–167.

6. "Japan in Depth: Abe's Yasukuni Visit a Personal, Calculated Risk," *Yomiuri Shimbun*, December 27, 2013, http://the-japan-news.com/news /article/0000900964.

7. Ibid.

8. Edward Wong, "No Meeting with Leader of Japan, Chinese Say," *The New York Times*, December 30, 2013, http://www.nytimes.com/2013/12/31 /world/asia/chinese-refuse-to-meet-japans-premier.html?_r=0.

9. "Park Slams Japan for 'Digging Up Past Wounds' after Abe's Shrine Visit," *Yonhap News Agency*, December 30, 2013, http://english.yonhapnews.co .kr/news/2013/12/30/37/0200000000AEN20131230004951315F.html.

10. Yoshihiro Makine, "Abe's Shrine Visit Blew Japan-S. Korea Efforts for Summit Sky-High," *Asahi Shimbun*, January 28, 2014.

11. Embassy of the United States of America, Tokyo, Japan, "Statement on Prime Minister Abe's December 26 Visit to Yasukuni Shrine," December 26, 2013, http://japan.usembassy.gov/e/p/tp-20131226-01.html.

12. US Department of Defense, "Joint Statement of the Japan, Republic of Korea, United States Defense Ministerial Talks," June 1, 2013, http://www.defense .gov/releases/release.aspx?releaseid=16054.

13. Ministry of Defense, Japan, "Defense of Japan: White Paper," http://www .mod.go.jp/e/publ/w_paper/2013.html.

14. The term "rear area" refers to Japan's territorial waters and international waters surrounding Japan (including the exclusive economic zone) in which no combat operations are conducted at that time and no combat operations are expected to be conducted throughout the period when the rear activities are carried out, and the space over these international waters. See: http://www .mod.go.jp/e/publ/w_paper/2013.html.

15. "Report of the Advisory Panel on Reconstruction of the Legal Basis for Security," Office of the Prime Minister of Japan, June 24, 2008, http://www .kantei.go.jp/jp/singi/anzenhosyou/report.pdf.

16. Yuichi Hosoya, "Japan's Plans for Collective Self-Defense," *Global Asia* 8, no. 4 (2013), 46–48.

17. "Talks Start with US on New Defense Plan," *The Japan Times*, January 18, 2013, http://www.japantimes.co.jp/news/2013/01/18/national/talks-start-with -u-s-on-new-defense-plan/.

18. "Japan, South Korea Discuss North Korea Concerns, 'Collective-Self Defense,'" *The Japan Times*, November 13, 2013, http://www.japantimes.co.jp /news/2013/11/13/national/japan-south-korea-discuss-north-korea-concerns -collective-self-defense/.

19. Park Young-June, "Japan's Assertion of the Right of Collective Self-Defense and Policy Recommendations for South Korea," *East Asia Institute*, November 19, 2013, http://www.eai.or.kr/type/panelView.asp? bytag=p&code=eng_multimedia&idx=12659&page=1.

20. J. Berkshire Miller, "Battle-Ready Japan?" *Foreign Affairs*, January 10, 2014, http://www.foreignaffairs.com/articles/140646/j-berkshire-miller/battle -ready-japan.

21. Park, "Japan's Assertion of the Right of Collective Self-Defense."

22. Bong, "ROK and US Views on the Foreign Policy of the Abe Administration."

23. Robert Dujarric, "Trying to Mitigate Japan's History Dilemma," *Asia Times Online*, October 30, 2013, http://www.atimes.com/atimes/Japan/JAP-01 -301013.html.

24. Sukjoon Yoon, "The New Chairman of the JCS and South Korea's Evolving Military Strategy," *PacNet*, no. 73, Pacific Forum CSIS, October 3, 2013, http://csis.org/files/publication/Pac1373.pdf.

25. "Japan's Push for Collective Self-Defense Stirs Dispute," *The Korea Herald*, October 29, 2014, http://www.koreaherald.com/common_prog/newsprint.php? ud=20131029000914&dt=2.

26. Ibid.

27. Karl Friedhoff, "Public Opinion on Japan Following Abe's Yasukuni Visit." The Asan Institute for Policy Studies, January 9, 2014, http://en.asaninst.org /public-opinion-on-japan-following-abes-yasukuni-visit/.

CHAPTER 7

Korea-Japan Relations under Deep Stress

Park Cheol Hee

Part 1: Korea-Japan Relations under Deep Stress

As is well known, Abe Shinzo and Park Geun-hye have not held a formal summit in the roughly one year since they took office and have no plans to do so. It has long been the case that new Korean as well as Japanese leaders call on each other upon assuming power. While many have pointed to reasons why this has not happened, the deeper meaning of this change in the bilateral relationship is worth further analysis. By offering some ideas about what to make of this transformation in bilateral relations with wide ramifications for security in Northeast Asia, I want to stimulate an exchange of views over the first part of 2014. My argument is that a crossroads has been reached, when issues that long festered cannot be avoided and choices that previously were seen as unnecessary or best left to the distant future are now a matter of immediacy.

Despite overlapping values and shared interests between the two countries, something serious has occurred in their relationship, which extends beyond the two leaders in charge. One clue is the negative legacy from the previous administrations. Following a December 2011 summit between Lee Myung-bak and Noda Yoshihiko, a sharp exchange of words regarding "comfort women" issues resonated in both countries in a manner that belied the previous period of optimism that Lee was the best hope for improved bilateral relations in decades and Noda similarly represented a golden chance for a leap forward in relations. Supposedly, a pragmatic leader of the DPJ who had abandoned progressive positions while drawing closer to the United States and a conservative Korean with the "best relations" ever with the United States were poised to improve the atmosphere for relations, even to reach a breakthrough.

Noda repeated the position that the "comfort women" issue had been completely resolved in 1965, when the two countries normalized diplomatic ties only to find that not only is the Korean perspective different, it is also impatient. It starts from the point of view that the issue had not been duly discussed at the time of the treaty between the two in 1965. Moreover, it recognizes that the issue had emerged in the early 1990s. That is the reason why Japan issued the Kono statement and Murayama declaration. The so-called Asia Women's Fund established by Japan did not make full closure of the circle, as the Korean women who had been victimized refused to accept compensation through the fund. The result of the summit was that a series of tit for tats on how to resolve the issue continued between Korea and Japan in 2012 without reaching a final solution. Even after new leadership in both countries, the foreign ministries failed to find a compromise solution. Accordingly, how to deal with the "comfort women" issue remains a lingering dispute between Korea and Japan. We need to look for a deeper explanation for why a compromise was not reached to show each side that progress could be made even if there would be no definitive meeting of the minds.

In addition to the "comfort women" issue, Lee's Dokdo (Takeshima) visit in August 2012 sparked an unexpected conflagration in Japan. Lee's careless comment regarding the Japanese emperor was like pouring oil on fire. After China had taken an assertive stance regarding the Senkaku (Diaoyu) islands in September 2010, territorial awareness among ordinary Japanese had been rising. In the eyes of the Japanese, Dmitry Medvedev's northern island (South Kuril) visit, Lee's visit, and China's aggression are all part of the same chain of events. Though three territorial issues with neighboring countries have different causes, together they have come to represent contempt for the declining status of Japan in international society. Abe, a candidate for the leadership of the LDP, took advantage of this uneasy feeling among the Japanese public to win the race and to lead his party to a resounding victory in the December elections.

If the "comfort women" and territorial issues are negative legacies from the previous regime, which were created neither by Abe nor by Park, the historical awareness issue is of Abe's own making. During the campaign for LDP leadership, Abe raised a hawkish conservative agenda, which many regard as including a call for constitutional revision, insistence on the necessity of collective self-defense, promises to revise the Murayama declaration and negate the Kono statement, and assurances that Japan would abandon its "masochistic," self-deprecating, historical perspective. He left no doubt of his right-wing political convictions. Though his comments have been somewhat toned down since he became prime minister, he remains a controversial figure in the eyes of Koreans, who closely follow his every move. In an extreme direction, Abe took a picture on a SDF fighter plane where "731" was written. He commented on the possibility of revising the Murayama statement, and he sent a gift to the Yasukuni Shrine long before going there. These were interpreted as right-wing signals.

In its first year, the Abe cabinet tried to show a reserved stance toward historical controversies as well. After Abe in July secured a stable political tenure following his winning a majority in the Upper House elections, he refrained from officially commenting on controversial issues. He abstained from going to the Yasukuni Shrine on August 15. In addition, Abe actively floated the idea of holding summit talks with Korea and China. He often said that the door to dialogue is always open.

Despite such caution, Park remained suspicious of his sincerity to deal with issues steeped in history. In the background of her reserved attitude were the following thoughts. First, she was still suspicious of the possibility of Abe visiting the Yasukuni Shrine. Also she could not figure out whether Abe would take a forthcoming attitude toward issues of concern, such as the "comfort women" issue. That is why Park repeatedly asked Japanese leaders to take a bold initiative to address historical controversies. Second, Park did not want to hold a dialogue for dialogue's sake. She had serious concern that failed summit meetings would produce irreparable damage to the bilateral relationship. In fact, Roh Moo-hyun's talk with Koizumi in 2005 and Lee's talk with Noda in 2011 led to more frictions rather than to any improvement in relations. In the hope of holding a truly successful summit meeting, she did not want it to be hastily arranged. These are the reasons why the Korean government delayed its response to the suggestion of summit talks.

Many observers of Korea-Japan relations expected that Abe and Park would meet on the occasion of at least one multilateral forum, such as the G-8, APEC, ASEAN+3, or the East Asian Summit. After no meeting materialized, the United States began tilting to the idea that Park was to blame for an outcome deemed to be bad for trilateral security cooperation. Faced with strong American pressure for a dialogue between Korea and Japan, the Korean side grew quite defensive from mid-October. Intellectuals and editorial writers in Korea also pushed for a Korea-Japan summit meeting at the earliest convenience. From around mid-November, Korean authorities were looking for some rationale to proceed on a pathway toward improving bilateral ties. It seemed as if a window of opportunity would soon be identified for breaking the stalemate between the two sides.

Given the new atmosphere, why then did Abe on December 26 suddenly visit the Yasukuni Shrine, which he knew would reverse any momentum? The response of Korea, as of China, was fully within his range of expectations. However, critical comments by the United States may have been an unexpected surprise to him. The US Embassy in Tokyo issued a statement of disappointment, which was an unusual diplomatic response. China took advantage of this situation as an opportunity to assail Japan and its diplomacy. Korea was genuinely disappointed and criticized Abe's Yasukuni Shrine visit, but it did not jump on the diplomatic bandwagon to criticize Abe from the perspective of international society. Though it closed windows of dialogues at the highest levels, it still kept low-level diplomatic dialogue open.

The situation has changed dramatically at the start of 2014. The Korean government is seen as more prudent and reasonably cautious in waiting for Abe to make a conciliatory move that few doubt he is obliged to undertake. The ball is in the Japanese court now. As Abe ruined the process of preparing for the summit by himself, he should also provide momentum for reviving the opportunity. The US government is placing the onus on Abe, leaving no doubt that it is wary about his coming moves and is determined to see him find a way forward with Park. Even many in Japan, who had abandoned hope in Park, are rethinking their allocation of blame. Whether these new circumstances will suffice to get Abe to become more serious is uncertain. After all, he let pass the late 2013 possibility of jointly searching for a way forward.

A few things can be done to move forward. First, Abe's Yasukuni Shrine visit should not be repeated. Paying a visit to the Yasukuni Shrine, where 14 A-class war criminals are enshrined, is considered an act of beautifying and glorifying the past war. Some indication that Park would not be embarrassed in this manner is important. Also, these visits contain a whiff of negating the America-endorsed postwar order, because it is the United States that is seen as behind the death penalties to war criminals. Second, the Abe cabinet would be well advised to show a sympathetic attitude to resolve the "comfort women" issue, while the victims are still alive. Time is rapidly running out. This should be resolved from the perspective of showing respect for human rights.

Third, escalating propaganda on the territorial issue should be avoided. In response to the Japanese initiative, the Korean government should be ready to deal with compensation for the forcefully mobilized workers during the colonial period. As an extension of past practices, it should continue to compensate the unidentified victims with full respect for the existing special laws in Korea. Furthermore, Korean should consider the possibility of widening defense cooperation between Korea and Japan in the near future.

For both sides, these steps will not be easy. Abe would have to take three initiatives at odds with what he has long advocated. The fact that in 2006 he took such an initiative on the Yasukuni Shrine suggests that there is a precedent, although his base has been aroused more than before and might see a betrayal. Park would also face an aroused public unprepared to let Japan off the hook on the forced labor issue and encouraged by the media to see defense cooperation as fueling the revival of Japanese militarism. To get Abe to show statesmanship of this sort and to help Park to make the case to the Korean people on defense, US leadership is likely to be needed. Indeed, a timely, quiet nudge by the ally of both states is anticipated.

Part 2: Japan-South Korea Relations under Deep Stress

As to the future of Korea-Japan relations, there are two contrasting perspectives. One is gloomy pessimism that foresees not only no easy solution to the pending issues, but even a further downward slide into a collision course between the two countries. Abe would not step away from his firm conviction that no further apology is necessary, according to this perspective, because he belongs to a circle of hawkish ideologues that have no intention of compromising in response to pressure tactics by Japan's neighbor. Though strategic interests would presumably lead Abe to appreciate the need to find a compromise solution on thorny historical controversies, he is likely to place more priority on domestic politics and comfort to his right-wing support groups. Also, in this perspective, Park, who takes a principled position on history-related issues, would not step back from her insistence that Abe show a sincere attitude toward issues stemming from the unhappy historical experience between Korea and Japan, including the "comfort women" issue. In other words, two principled political leaders, who give priority to domestic politics rather than to regional strategic ties, have little incentive to move forward toward reaching a breakthrough. This is what most observers seem to expect.

The other perspective takes a cautiously optimistic view. Grounds for optimism do not stem from a shared strategic conception about the future of the region, though. Rather, cautious optimism arises from the expected mutual loss from noncooperation. If the two countries continue to muddle through, they have to face the year 2015, the fiftieth anniversary of normalized ties, with embarrassingly complicated issues related to historical controversies. Bilateral relations may be left in such an unmanageable state that full recovery would be impossible in the foreseeable future. Both Korea and Japan have incentives to avoid this worst-case scenario. After all, troubled ties between the two countries would only work to the benefit of China and North Korea, while deeply undermining US strategic interests and shaking its trust in what have been strong alliance partners. Hence, Washington is likely to make the utmost effort to draw its two allies closer. According to this view of an outside-assisted recovery of cooperative ties, Korea and Japan may not feel comfortable about resuming normal relations, but they would have insufficient reason to avoid restoration of at least the trappings of business as usual. The two countries would be pulled, however reluctantly, by external circumstances and by an indispensable third party to a state of cooperative partnership.

The United States already has begun working toward the second scenario. John Kerry visited Seoul on February 13 and made it clear that Korea-Japan ties should be improved before Obama's planned trip to Asia in April. Apparently, the United States accepted a reminder from Korea that a presidential visit to Tokyo without calling on Seoul at this time would send an unexpected diplomatic signal that it appreciates Japan more, while being ready to sacrifice Korea and leaving Japan with the misleading impression that the United States does not care about the historical controversies between the two countries. It could even be interpreted as condoning Abe's Yasukuni Shrine visit that the United States had opposed. Thus, the White House decision to shorten the presidential visit to Japan and add a stop in Korea suggests a balanced approach. It leaves the two US allies starting on an equal footing in terms of the US commitment, while adding an element of urgency and making sure that the visit proceeds successfully as viewed from the angle of boosting trilateral cooperation.

The Obama administration is not hesitant about taking a position on Japan's strategic choices. It is highly supportive of Japan's move to make proactive contributions to international as well as regional security. Also, it does not oppose the idea of making Japan a normal military power if the Japanese people choose to do so, including introducing the right of collective self-defense or even constitutional revision. Japan's security activism can be helpful for the United States. However, the Obama administration draws the line on any Japanese moves that intentionally provoke its close neighbors with historical and territorial controversies. Last October, John Kerry and Chuck Hagel visited Chidorigafuchi National Cemetery in the hope of conveying a message to Abe not to visit the Yasukuni Shrine. Also, Joe Biden made it clear that he is not supportive of an Abe visit to the shrine. That is the reason why the United States expressed deep disappointment about the visit. Despite repeated lobbying from the Japanese, the US Congress is not necessarily against the idea that Japan should apologize about the "comfort women" issue. Positively encouraging Japan's security activism and proactive contribution to global peace, but unequivocally discouraging history revisionism advocated by right wingers in Japan, the White House has made its position clear to its allies.

The US role is pivotal for getting its two allies to draw closer; however, such external pressure only serves as a convenient excuse for short-term collaboration rather than eliciting durable cooperative ties between Japan and South Korea. Without clear self-awareness about why cooperation serves the vital interests of their own country, Korea and Japan might only show smiling faces in front of a common ally while still remaining mutually suspicious of each other. Even if reluctant cooperation is better than no cooperation, more than face-saving measures are needed to go forward to enhanced strategic cooperation.

For the purpose of eliciting enhanced cooperation from Korea, Japan has to show a nuanced position to its neighbor. First, Japan had better develop a sophisticated eye for differentiating Korea from China. Both espouse what Japanese regard to be "anti-Japanese" sentiments. However, unlike China, such feelings among the Koreans are hardly translated into political action, as in the form of demonstrating against Japan. Though the mass media depict Japan in pejorative terms, the ordinary public still enjoy Japanese popular culture—cuisine, songs, and dramas, as well as novels. People-to-people relations between the two nations remain resilient. Economic transactions are not immediately affected by the politically turbulent relationship. There is an erroneous media-created image that treats Korea as a part of China's sphere of influence. Japanese hawkish conservatives tend to put Korea and China in the same basket, even treating the two as if they are twins, despite enormous differences in their political systems, lifestyles, behavioral orientations, and strategic calculations. The fact that both protest against Japanese provocations related to historical controversies may make many Japanese think that they are the same; however, if Japan abstained from taking an ultraconservative position on such issues, Korea would not associate itself naturally with China. In other words, Japan can dissociate Korea from China if it takes history issues off the table.

Second, Japan had better take a forthcoming attitude in dealing with rising China. It is understandable that Japan is wary of rising China's assertive external strategy, especially in the maritime arena in the East China Sea as well as in the South China Sea, but antagonizing China may not serve Japan's interest, because it may only escalate tensions between the two. Building a strong wall between Japan and China may serve the short-term security interests of Japan, but it is more likely to lead to a security dilemma in the region in the long run. As Japan demonstrates its will to promote "proactive pacifism" on the global scene, it can actively show its will to promote peace and stability in Northeast Asia. This would indirectly serve the purpose of improving Korea-Japan ties. Korea does not want to be torn between the two regional powers. If the two powers act in a centripetal way in a regional community, then Korea would welcome playing a bridging role. Of course, this depends on China as well as on Japan, and Koreans are not expecting one-sided abandonment of principled positions.

In order to upgrade cooperative ties with Japan, Korea should also take reasonable and responsible positions. First, Korea should be much more appreciative of Japan's legitimate actions in the past. Korea is too preoccupied with the unhappy Japanese colonial rule and always recalls those experiences first. This leads to twisted understanding even about postwar Japan. Korea has every reason to be cautious about any suspicious actions and words by Japanese politicians, but it should focus its perspective on the newly evolving regional configuration to avoid a diplomatic impasse with Japan. History can never be forgotten, but it can be forgiven. Second, Korea should understand the geostrategic importance of its location in a changing regional power transition. While it can gain considerable economic advantage by maintaining friendly ties with China, in terms of security, South Korea has no choice but to stand with the United States and Japan, at least until it unifies with North Korea. At a minimum, it should maintain balanced diplomatic ties with Japan and China. Taking a lopsided position against either will make Korea insecure and unreliable.

Part 3: Korea-Japan Relations after the Hague Summit Meeting

President Park Geun-hye and Prime Minister Abe Shinzo finally met at The Hague, Netherlands, on March 25, 2014. President Obama took pains to arrange this trilateral summit meeting on the fringes of the nuclear security summit. The three leaders discussed an issue of common concern, North Korea. As the first meeting between Abe and Park after assuming power, respectively in December 2012 and February 2013, and between the leaders of the two countries since May 2012, it left something to be desired. Park and Abe refused to sit side by side. Obama sat in the middle. Though Abe said hello to Park in Korean, she did not favorably respond or smile in return. As the meeting was organized by Obama at the American embassy, the leaders concentrated on common security challenges—North Korea's nuclear development and missile launches—, agreeing that North Korea should be denuclearized. Before the Six-Party Talks are convened, the three sides have agreed to hold a "security consultation meeting" to make sure that common concerns are coordinated.

The Hague meeting was a successful start in that two reluctant leaders had a face-to-face meeting after hard US pressure on the Abe cabinet to make this possible. After Abe's Yasukuni Shrine visit, the US open expression of disappointment was followed by growing alarm about his cabinet's attempt to revise the Kono statement, by which Japan sincerely apologized for the "comfort women" issue after independent investigation by the Japanese government. Given cautious US encouragement of a change in attitude, Abe made it clear on March 14 that he would not pursue revision of the Kono statement. The Blue House, which remained highly suspicious of Abe's sincerity on this issue, appreciated Abe's remarks. Also, the US State Department welcomed this development.

On March 12, two days before Abe's announcement, administrative vice foreign minister Saiki Akitaka came to Seoul on the pretext that he would meet the newly appointed Korean vice foreign minister Cho Taeyong, knowing that the Korean government was repeatedly asking for "sincere action" on issues of deepest concern to the Korean people. Saiki returned to Tokyo without even staying the night in Seoul. He had a one-on-one meeting with Abe the next day to report on his trip. Realizing that the Kono statement should not be revised if the three-way summit meeting were to go forward, Abe made his remarks the following day.

The United States Standing between Korea and Japan

As Abe made numerous promises during the LDP presidential election campaign in 2012 regarded as part of a right-wing agenda—revising the Constitution, visiting the Yasukuni Shrine, revising the Kono statement, and so on—the United States, not only Korea and China, cast a suspicious eye on his political ideology. When Park was inaugurated, Abe totally failed to grasp the chance to ameliorate relations between Korea and Japan. Vice Premier Aso Taro, Abe's envoy to the inauguration ceremony, ruined the first contacts between the two new administrations. Instead of delivering a congratulatory message, he engaged in a history debate with Park on the first day of her official work, saying, "In the United States they had a Civil War between the South and the North. People living in the North and South still have different historical memories and different historical perspectives. How can two countries like Korea and Japan share similar historical perceptions when even people living in the same country hold different perspectives?" This remark made Park and her close aides furious. After returning to Tokyo, Aso visited the Yasukuni Shrine in April, together with three other cabinet members. More than 160 LDP members followed suit. Aso had a slip of the tongue again when he suggested that Japan could learn from the Nazi approach to constitutional revisions when it tries to revise its peace constitution. In April 2013, Hashimoto Toru, the Osaka major and a leader of a new opposition party, *Ishin no Kai*, stated that "comfort women" were available everywhere and Japan alone should not be blamed. In the same month, Abe inadvertently raised his thumb on a SDF fighter plane where the number 731 was written. All these moves were perceived as a right-wing turn in Japanese political circles. Korea and China showed deep anger at these reckless moves. It was not only neighboring countries but also the United States that was frustrated by these relentless remarks and actions. Thinking that Abe was blocking progress between Japan and Korea, on the one hand, and Japan and China, on the other hand, it began putting pressure on the Abe cabinet to show restraint. Abe toned down his political rhetoric, eventually. He did not visit the Yasukuni Shrine on August 15, although he sent gifts to the shrine.

After refraining from the shine visit, Abe repeatedly sent signals to Korea that the door is open to hold a summit meeting at the earliest convenience. During September and October, there were multilateral summit meetings—the G-20, the UN General Assembly meeting, APEC, ASEAN+3, and the East Asian Summit meeting. However, Park did not show any sympathy with Abe's proposal. On the contrary, she suggested several times during her overseas trips that Japan should be forthcoming in showing sincere apology over its history issues and urged its leader to resolve the "comfort women" issue. Japanese public opinion turned more negative and retorted that the Korean attitude of blaming Japan in third countries could hardly be accepted. From around October, Abe began actively to advance a security agenda—establishing a National Security Council (NSC), changing the constitutional interpretation in a way that recognizes the right of collective self-defense, and devising a National Security Strategy (NSS). In Korea, the mass media covered these moves as if Japan was planning to become a military superpower or was reverting to the prewar order. Even though opinions were divided in Korea, the mass media led a campaign not to allow Japan's introduction of the right of collective self-defense. This negative depiction of Japan's moves sent an unforgivable message to the United States, considering that this would make possible a change in the US-Japan alliance into something truly reciprocal. Also, Japan would be able to help US forces in the Pacific, enlarging the scope of possible military action. From the perspective of the United States, there are few reasons to oppose the introduction of the right of collective self-defense. In this connection, it began to look at the Park administration as if it was trying to get in the way of a US security alliance. It was around that time that the mood in Washington began turning against the Park administration. Pundits in Washington argued that Park was mainly responsible for the failing relations between Korea and Japan, because she was too rigid and principled. They also claimed that Korea has been too preoccupied with the history issues. Washington began pressuring the Park administration to ameliorate bilateral relation between the two countries.

Reflecting US concerns and repeated proposals by Abe, from around mid-November, the Korean government opened a diplomatic channel to explore the possibility of a summit meeting in the not-so-distant future. Though negotiations were not publicized, both Korea and Japan seriously sought a meeting at the end of the year or early in 2014. However, to everyone's surprise, Abe made up his mind to visit the Yasukuni Shrine on December 26. Only a few close aides were aware of his determination, some of whom tried to stop him from going until the last moment. This visit ended the possibility of a meeting with Park. December 26 was the one-year anniversary of his inauguration, and he gave priority to his domestic political audience rather than to diplomatic considerations. The ball was back in Japan's court. People in and out of Japan realized that Abe was primarily responsible for the deteriorating relationship. This was a shock to Washington as well. As a warning against a Yasukuni Shrine visit by Abe, Secretary Kerry and Secretary Hagel, when they went to Tokyo for the 2 + 2 meeting in October 2013, visited the Chidorigafuchi National Cemetery, a memorial site for unknown soldiers. In November 2013, when Vice President Biden was making the rounds in Japan, Korea, and China, he made sure to warn Abe that the Yasukuni Shrine should be avoided. Hence, there was another turning point in dealing with Korea and Japan on the part of the United States. Without Abe's Yasukuni Shrine visit, a summit meeting between Park and Abe at the Davos Forum could have been a possibility. Obama was scheduled to visit Japan and Korea in April 2014. His efforts to forge a trilateral partnership were in jeopardy. Also, Korea and Japan felt pressure to make Obama's East Asia visit a successful one. This way of thinking made the two leaders respond positively to the US proposal of getting together in The Hague. The US role was crucial in drawing two reluctant partners closer. From now on, whether Korea and Japan can make progress toward summit meetings without US mediation is in doubt.

Colliding National Strategic Identities as a Hurdle

Korea and Japan are not necessarily trying to get away from each other. Both of them know that cooperation is a better option for them than muddling through or remaining aloof. The two countries share similar regime qualities, including democracy and a market system. Both countries respect human rights, freedom of speech, and the rule of law. The two are vital US allies and forces for maintaining the liberal international order in East Asia, which is led by the United States. Hence cooperation seems to be the naturally expected outcome.

When it comes to subjective dimensions, however, the two have dissimilar perceptions and different priorities in handling bilateral and external affairs in general. Japan, alarmed by a rising and increasingly assertive China, would like to balance against it. In Japan, China is often perceived as a challenger and even as a potential enemy. How to cope with rising China is the fundamental concern of Abe's external strategy. He wants Korea to be a part of a balancing coalition. Accordingly, Abe puts emphasis on future-oriented security and economic cooperation with Korea. He wants to put aside or, if possible, bypass historical controversies between the two countries. For him, dialogue between the two should be conducted mainly on the topic of security cooperation, especially in the context of rising China. But Park's first agenda item is not security cooperation. She inherited a negative legacy, from both Lee Myung-bak and Noda Yoshihiko. Making a breakthrough on the "comfort women" issue, which looms at the center of public opinion, is a foremost concern. Only 55 victims survive, and Koreans insist that this issue should be resolved from a universal human rights perspective. She has no intention of bypassing this issue. Abe shows either a very negative or a very ambiguous attitude toward it. Japan, which prioritizes security cooperation, and Korea, prioritizing the "comfort women," cannot take a step forward together.

At The Hague summit meeting and in its aftermath, the Korean and Japanese governments agreed to set up two meetings. One is a mechanism designed for furthering cooperation among the three countries on common security challenges, above all, the North Korean security threat, including nuclear development and missile launches. This security consultation mechanism is to reflect shared concerns through prior consultations before the Six-Party Talks are convened. In this context, Korea and Japan are likely to advance steps toward security cooperation. This consultative body among defense officials partly reflects Japan's priority concern. Another step is for both countries to launch at the director-general level talks on issues of immediate concern, including the "comfort women" issue. As a gesture toward this, on March 25, Japanese foreign ministry officials visited the house for surviving "comfort women," the "House for Sharing" (*nanum eui jip*). The fact that the two sides plan to sit together at the table to discuss this troubled issue is a good sign. In addition, leaders of the Korea-Japan Parliamentary League met on March 25, to discuss the role of politicians in enhancing channels of communication. The security consultation mechanism reflects Japanese concern while the director-general's dialogue reflects Korea's priority concern. Using these two tracks, the two sides will explore chances for improving bilateral relations. It remains to be seen whether these bureaucratic channels of communication will bear fruit. The two states emphasize different issues, deriving from divergent conceptions of national strategic identities, at the bottom of which are colliding or at least mismatched priorities.

Abe wants to make Japan economically vibrant and militarily flexible. In order to cope with a rising China, his cabinet is striving for a multilayered strategy. Strengthening the US-Japan alliance stands as the top priority, essential since Japan can hardly deal with China on its own. As a part of this endeavor, Abe aims to change the constitutional interpretation in a way that permits Japan to exercise the right of collective self-defense. Also, Abe's Japan is ready to contribute more to international security under the banner of "proactive pacifism" ("proactive contribution toward peace"). Defending the liberal international order, which is mainly maintained by US leadership, is a shared strategic goal, in connection with which Japan wants to position itself as a status quo power, not a potential revisionist power, as China is portrayed. Abe repeatedly insists that international law should be observed and that any country that tries to change the existing maritime order by force should not be condoned. For this purpose, Japan would like to align with democratic maritime powers, namely, the United States, the United Kingdom, Australia, India, and ASEAN countries. In this strategic scheme, South Korea is mentioned from time to time, not always. Although it has a democratic regime and a clear interest in preserving the existing liberal international order, it is not usually designated as a close partner with which Japan should cooperate. This is a missing link in Abe's national strategic identity conception, stemming from a contradiction within it.

Another pillar of Abe's national strategic identity is advocacy of historical and territorial sovereignty. Unlike his predecessors, Abe does not want to sustain "kowtow diplomacy" based on apology, repentance, and compensation. Instead, he wants to make Japan a country with pride and confidence. In this national strategic conception, it is not strange at all that Abe's Japan stands strongly against Chinese and Korean requests for apologies related to historical transgressions. Already when he was prime minister in 2007, he led in a cabinet decision that the Kono statement be revised. Even though Japan's position is complicated by the fact that Japan virtually controls the Senkaku islands and Korea controls Dokdo (Takeshima), Japan claims that both are under Japanese sovereignty. If historical and territorial controversies are taken into account, Korea can hardly be regarded as a partner with which Abe's Japan could work closely. In the construction of Abe's national strategic identity, Korea's priority in Abe's external strategy is not high. It is a troubled partner for Japan. From the angle of national security strategy, Korea should be embraced more tightly, but, in light of historical and territorial controversies, Abe is dragged down by this troubled relationship. The Abe cabinet has not found a solution to this contradiction.

Park characterizes her diplomatic strategy as "trustpolitik," or trust diplomacy. Her primary objective is to reconstruct the twisted inter-Korean relationship. The "peace process on the Korean Peninsula" is the designated way to realize this goal. According to this scheme, abnormal behavior should not be respected but corrected. Upon returning to normal, acceptable, common sense behavior, South and North

Korea can undertake joint efforts to build trust, starting with functional and humanitarian areas. Park remained principled when North Korea threatened with harsh remarks and daunting threats in the spring of 2013. Together with the United States, she took strong initiatives to show her will to strike back. Later North Korea warned that it would close the Kaesong Industrial Complex and did so, but Park waited for the situation to stabilize on the assumption that North Korea would have more to lose. Bringing back normal and acceptable behavior is the starting point of the trust-building process. This concept was applied to the Northeast Asian situation as well. According to her cabinet, what is lacking is trust in Northeast Asia. She refers to the Northeast Asian reality, where distrust and conflicts are deepening in the middle of widening economic and human exchanges, as the Asian paradox, at the heart of which lies abnormally behaving Japan under Abe. As Abe's Japan shows ambiguity about the past and has no intention to further apologize while glorifying historical transgressions, the Asian paradox does not draw to an end. Hence, reviving normal behavior is necessary to build trust among Asian neighbors too. Forgotten here is aggressive behavior by China. Because of this strategic scheme, Japan is depicted as at the center of worsening ties among Asian countries.

Park also suggests that Korea should take the lead in molding peace and cooperation in Northeast Asia. For her, the black hole for Northeast Asia is North Korea. In order to realize peace and cooperation, Korea should work closely with the United States and China, two nations that have high stakes in North Korea. In order to manage provocative North Korean moves, on the one hand, and to maximize the chance that North Korea will turn to the outside world as a reform-oriented open regime, on the other, she considers cooperation with the United States and China to be necessary. In particular, China's cooperation is an indispensible part of this strategic scheme, because it has more leverage over the North Korean regime than any other country. In the process of reshaping the order on the Korean Peninsula, Japan is not likely to play an active and important role, at least in the beginning. Its preoccupation with the abductees issue limits its diplomatic scope. In order to establish a peaceful and cooperative Northeast Asia, Abe's remarks and behavior can hardly be accepted as they are. Rather, Park repeatedly has asked Abe to resolve to take care of the unfinished historical pain represented by the "comfort women" issue. This is not a precondition for her, but it is a cornerstone for further steps toward building trust between suspicious partners. Hence, for handling the North Korean issue, as well as for building peaceful and cooperation in Northeast Asia, Japan is not a priority. According to this strategic scheme, Abe's Japan should first get back to a normal status whereby it does not deny, distort, and glorify its troubled history. Japan is expected to act in accord with principles recognized in international society. Seen from this angle, it is not surprising at all that Japan's priority in Park's strategic scheme is not high.

Analysis of Abe and Park's national strategic identity conceptions tells us that the strategic weight of the counterpart is not high enough to give urgency to normalizing ties. At least, both political leaders do not want to shoulder the domestic political burden of reviving friendly ties with the other country. This is one of the reasons why they had to be pushed by Obama to meet in The Hague. Even after this summit, they have to recalibrate their strategic stance toward each other in order to satisfy both domestic and foreign audiences.

What Should Be Done?

If national strategic identities contradict each other or at least are a mismatch, what should be done in order to bring back normal friendly ties? There is a path, but it would be a very narrow one for each side to take careful steps forward, aware of the need for painstaking negotiations before they can reach a compromise solution. On the part of Japan, the first challenge is to keep Abe's promise, despite his personal ideological convictions, that he would not change the Kono statement. This means that he is willing to admit the element of coercion in mobilizing "comfort women" during the colonial period. Sticking to this position will serve as a barometer for the resumption of trust between Korea and Japan. Though Abe made this promise in the Diet, a special advisor to him, Hagiuda Koichi, asserted that if new evidence is found in the process of reviewing the Kono statement, it would not be strange at all to make a new political statement. This is not fully in line with Abe's previous assertion. Hagiuda's remark is based on the assumption that the Abe cabinet will continue to reinvestigate the process of making the Kono statement, as Chief Cabinet Secretary Suga Yoshihide mentioned. Education minister Shimomura Hakubun went further in saying that there is no unified political will to support the Kono statement. If this is the case, nondenial of the Kono statement may go hand in hand with reexamination of the Kono statement or even with the domestic backlash against Abe's political decision, confusing observers, including the Korean government. Maintaining a unified, consistent position regarding the Kono statement is likely to prove to be a political challenge for

Abe. It is in this context that the Abe cabinet's decision on April 1 neither to revise the Kono statement nor to draft a new political statement is a meaningful step forward.

Whether Abe's Japan can upgrade the strategic weight of Korea in national strategic identity is another challenge. In the Japanese media, Korea is often depicted as "little China." It tends to put Korea and China in the same basket and feature these two neighboring countries as troublesome partners for Japan. Of course, they are not the same, politically, economically, or strategically, in the context of Northeast Asia. Shared historical perceptions derived from memories of Japanese imperialism may be an exception. What is lacking in the mindset of the Japanese is consciousness that Korea is a divided nation. In order to reunify the nation, South Korea needs help, or at least recognition, from China, which shares a border with the Korean Peninsula. This geopolitical position makes the South Korean approach to China dissimilar to Japan's, but this does not mean that South Korea is in the embrace of China. Japan should recognize the simple fact that the Korean lifestyle resembles Japan's more than China's in every sense of the word. When it comes to security issues, Korea would not take sides with China. Cultural ties between Korea and Japan are so thick that dividing the two is unnatural. Considering all these elements, Japan should rediscover the strategic importance of Korea in the context of the shifting power configuration in East Asia. At this juncture, China well understands how important it is to bring Korea to its side. The United States also recognizes the geopolitical importance of Korea in a changing strategic milieu. Unfortunately, Japanese leaders are slow in catching up with this strategic game; they are too preoccupied with domestic politics.

Whether Japan can find a way to contribute to the unification led by South Korea is another challenge. Japan shows single-minded concentration on the abductees issue to score political points with domestic audiences. This is not something to blame; however, in the eyes of Koreans, playing the North Korean "card" looks like a diplomatic pressure tactic to entice Korea to the negotiating table with Japan. China under Xi began revealing a new perspective that unification, even under South Korean initiative, could be accepted. Although there may be a tactical element in this gesture, China may be recognized as a country that is sympathetic to the idea of reunification. If close coordination with Korea can be achieved, Japan can gain more from reunification of the two Koreas. Abe's Japan should widen its strategic vision to the entire peninsula and think about ways to open the way to reunification rather than simply gaining political points by addressing the abductees issue for Japan's own sake.

Korea should also do its utmost. First of all, it should deal with Japan-related issues in a comprehensive manner, not mired in historical issues alone. Historical consciousness comprises a very important part of bilateral talks and is a basis for trust building; however, Korea-Japan relations should be situated in a wider strategic context in Northeast Asia. Although it is undeniable that China's role is vitally important for handling North Korean affairs, Korea should maintain a balanced approach to jointly dealing with Japan and China. History issues can be a good entry point to further enhance ties between the two countries, but negotiations between them should not stop there.

Second, Korea should send a clear signal to Japan that historical controversies have an endpoint. As long as Japan is suspicious about Korea's intentions, historical controversies can be perceived as a hope-lessly endless game. If Japan is forthcoming with proposals for handling the "comfort women" issue and others, Korea should greet Japanese suggestions positively and try to strike a constructive deal. Negotiating with Japan is a reciprocal process. Unless the Korean government shows willingness and endurance to address issues of concern together, Japan will lose any temptation to resolve politically delicate issues.

Third, Korea is expecting to play a modest mediating role among regional powers. It is not in a position to become a leading nation in Northeast Asia, but under conditions of Japan and China being seriously confrontational, Korea can play a positive role of linking the two, facilitating communications between them and suggesting modest alternatives for handling thorny issues. With these goals in mind, Korea had better initiate the revival of the trilateral summit among Korea, Japan, and China in the nearest future.

Unless Korea and Japan successfully manage thorny historical controversies, 2015, which is the fiftieth anniversary of signing their normalization treaty, may turn out to be a disaster for both of them. They should strive for a constructive, future-oriented, strategic partnership on the basis of cherishing the wisdom of facing historical truth and acting humbly not to repeat past failures. Recognition of this is needed for building on the opportunity offered at The Hague for a new beginning.

PART III

Japan's Emerging National Security Policy

CHAPTER 8

Complacency and Indecisiveness in Japan's New One-Party Dominance and Foreign Policy

Koichi Nakano

Japanese politics has entered a new era of one-party dominance as the LDP-New Komeito coalition government led by Abe Shinzo secured a comfortable majority in both houses as a result of the coalition's handsome victory in the July 2013 Upper House elections. The so-called twisted parliament (in which the ruling parties lacked a majority in the second chamber) is now a thing of the past, and the government is in a position to pursue its own legislative agenda unobstructed by national elections for the next three years. Many now expect decisive action, including in foreign policy.

Abe continues to enjoy favorable polls, thanks to the early successes of "Abenomics" and is most recently given a further boost by the conclusive win of Tokyo's bid for the 2020 Olympics. Indeed, the new one-party dominance today is different from the "old" one-party dominance. Until 1993, the LDP ruled without any break since its founding in 1995, but it always faced the leftist opposition of the socialists and communists, who together had at least a third of the seats in the Diet. Today, the left has only 3 percent of the seats (even with the recent gains made by the Japan Communist Party).

The DPJ that grew to become the main alternative to the LDP at the expense of the left parties is now in tatters. None of the parties that splintered from it, including the People's Life Party of Ozawa Ichiro, won any seat in the July election. Your Party and Japan Restoration

Party (JRP) are both little more than satellite parties of the LDP, both in terms of membership and policy preferences. If anything, the fact that the JRP did not do better in the polls was a disappointment for Abe, who is in search of a greater alliance of political forces that are clearly in favor of revising the constitution.

The fact that there is no real opposition to speak of today is a historically unprecedented situation in postwar Japan, and that is why the new one-party dominance is significantly different from the old. Theoretically at least, the two-thirds majority in both houses of parliament that eluded the LDP throughout the postwar period is conceivably within its reach at long last. If one adds the seats occupied by the LDP, Komeito, Your Party, and JRP, there is just enough to go over the threshold even in the Upper House. Of course, a four-party agreement on the revision would be very hard to attain, but for the proponents of constitutional revision, prospects have never been better. Moreover, even if revision is hard to achieve in the short term, by placing the issue on the agenda, Abe and his followers can further drive a wedge in the already much weakened DPJ. In such a scenario, its right wing will simply form another satellite party of the LDP.

In short, the rightward drift of the party system in Japan that has been taking place for the past couple of decades continues, much to the advantage of the LDP, which is not only much larger than all the other parties, but also may get to choose between different coalition frameworks, and thus has potentially greater leverage over Komeito.

When one scratches the surface, however, it becomes clear that the overtowering position of the LDP is, to a considerable degree, an artificial creation of the electoral systems that grossly exaggerate the LDP's plurality of votes into a massive majority of seats. In the December 2012 Lower House elections that ended in the LDP triumph, the party was actively supported by a mere 16 percent of all eligible voters (including those who abstained) in the Proportional Representation (PR) segment—in fact, fewer votes than in its electoral rout of 2009 (18.1 percent of eligible voters) when the DPJ took power. Much the same can be said when one calculates the percentage of votes received by the LDP out of all the eligible voters in the PR contest in the upper house: 16 percent in 2007 (when Abe's first government was defeated soundly by the DPJ), 13.5 percent in 2010 (when the LDP came ahead of the DPJ in seats gained, even though it was trailing the DPJ in votes), and 17.7 percent in July 2013. In other words, the LDP cannot be said to be back in power due to a surge of popular support, as only slightly more than one in six Japanese voters actively vote for it, even today.

The power position of the LDP is inflated by the following four related factors. First, the non-PR component of the election in both the Lower and Upper Houses is notoriously skewed in favor of the rural vote, where the LDP has always been strong. The bias is so pronounced that even the conservative courts in Japan have been finding the current state of affairs unconstitutional in a series of rulings. Second, the DPJ support collapsed (from 22.4 percent of eligible voters in 2007 to 6.8 percent in 2013 in the PR section of the Upper House) as the party was discredited for its poor record in office. Third, with the split and decline of the DPJ, the party system became highly fragmented, and thus split the anti-LDP vote to the LDP's great advantage. Fourth, facing a confusing array of new parties when neither of the two major parties enthused a large number of the electorate, the voting turnout declined sharply—the lowest turnout in the postwar era in December 2012 and the third lowest in July 2013, respectively. It is important to realize that the current parliamentary strengths of the government are grossly exaggerated because it accounts for both its complacency and indecisiveness in the first few weeks following the Upper House election victory.

The government has reason to feel complacent because it is now very literally the only game in town. It is also fully aware that the Japanese media realizes this, and that the journalists will have to think twice before they get overly critical of it. Abe's grip on the media was shown by the latter's lack of any real teeth in covering the leakage of radioactive water in and around the Fukushima site, and the premier's personal, and obviously untruthful, reassurance that the situation was under control. Clearly if Abe can get away with that, he should be able to get away with a lot of things.

The government also has reason to be indecisive, however, because the return of support for the LDP, in spite of the near total collapse of support for the DPJ and its splinter groups, and in spite of less than overwhelming support for the JRP or Your Party, has been minimal in terms of actual votes, and the little that has been gained seems largely to derive from hope and expectations for "Abenomics"—not for the reactionary, nationalist policies that are dear to Abe and his colleagues. Abe realizes that he stands on rather fragile ground in reality, and one small error can potentially quickly lead to a sharp decline in the cabinet support levels in the polls.

Indeed, the combination of complacency and indecisiveness was behind Abe's decision to put off the customary and much anticipated postelection cabinet reshuffle this summer. One may say that Abe's position was secure enough to allow him to keep his cabinet unchanged even though several of his ministers (including the gaffe-prone deputy premier Aso Taro) were already widely noted to be less than competent. It could also be said, however, that he was afraid to lose his authority by going ahead with the reshuffle as it would provided ample opportunities to disappoint his colleagues in the party who would be either demoted or not promoted, while risking the revelation of scandals of those who would be appointed to join the cabinet anew. It was easiest not to touch the cabinet, even though it consists of many second-class players, with the noted exception of Chief Cabinet Secretary Suga Yoshihide, who is really the power behind the throne. Thus, while the rightward shift of Japanese politics that has been taking place in fits and starts over the past two decades continues with the emergence of a new one-party dominance of the LDP, complacency and indecisiveness are likely to be keywords in understanding Abe's policies for the foreseeable future.

In Japan's relations with its Asian neighbors, a deliberate and decisive escalation of tensions with China and South Korea over territorial and history issues is unlikely, but at the same time, a complacent attitude among policymakers appears widespread that the Chinese and the Koreans should now realize that they have to soften their attitudes since Abe is here to stay. It is hard to imagine serious efforts to improve ties initiated by the Japanese side, and not surprisingly, all that took place at the G-20 in early September was brief chit-chats between Abe and Xi Jinping and Park Geun-hye, respectively.

Similarly, complacency and indecisiveness marked the start of US-Japan ties under Abe so far. Harshly critical of the DPJ's steering of the alliance, Abe used to say it was time that the LDP returned to power to rebuild ties with the United States that were nearly destroyed by the DPJ, but instead of receiving a warm hug from Barack Obama, Abe's complacency has been badly shaken by a chilly reception so far. It was only thanks to Syria that Abe got to meet Obama for the second time in St. Petersburg. What Abe does not seem to realize is how times have changed since the Koizumi-Bush "honeymoon." In the past, the United States used to overlook, even encourage, the rise of "healthy" nationalism in Japan insofar as that was the price to be paid for such reformist leaders as Nakasone Yasuhiro, Hashimoto Ryutaro, and Koizumi Junichiro to open the Japanese market to American corporations and to have the Self-Defense Forces do more and get further integrated with American forces in the context of US military transformation. It bears reminding ourselves that all three of these reformist leaders visited Yasukuni Shrine to placate the nationalists within the LDP, to which Washington turned a blind eye.

China has since overtaken Japan as the number two economy in the world, while relations took a turn for the worse over the Senkaku/Diaoyu islands, thus today, the nightmare scenario for the United States is to be drawn into a military conflict between Japan and China. Abe's nationalist beliefs are seen as potentially highly dangerous for US interests in Asia. In fact, Japan's inability to improve ties with South Korea, the other key US ally in Northeast Asia, over the "comfort women" issue in particular, seems also to be a source of great frustration for Obama's policy advisors.

Thus, no matter how hard Abe tries to resort to the past magic formula—neoliberal economic reform and nationalism—in dealing with the United States, by joining the Trans-Pacific Partnership negotiations and by exploring further deregulation of the Japanese economy as part of "Abenomics," more than any opposition party inside Japan, the US government is acting, at least for the time being, as the most effective brake on his ambitions to revise the Kono and Murayama statements, visit the Yasukuni Shrine, and rush into a revision of the constitution, thereby preventing him from adopting a more decisively nationalistic line of foreign and security policy. With the new one-party dominance that resulted from the recent elections, unstable government and opposition gridlock will no longer be the curse of Japanese political leaders, but complacency and indecisiveness may well become the Achilles' heel of the Abe government.

CHAPTER 9

Evolution of Japan's National Security Policy under the Abe Administration

Yamaguchi Noboru

Since the latter half of 2013, the administration led by Prime Minister Abe Shinzo has been developing a national security policy characterized by the following six points: (1) a systematic approach toward national security policy; (2) a National Security Strategy (NSS) with Proactive Contribution to Peace as its key concept; (3) a Dynamic Joint Defense Force as the goal in the defense buildup; (4) a seamless response to "gray zone" events as an operational concept; (5) reconstruction of the legal basis for security; and (6) serious consideration of the trends in the regional security environment represented by the US rebalance toward the Asia-Pacific region and China's rise.

A Systematic Approach toward Security Policy

The Abe administration initiated a systematic approach toward national security policy, creating a mechanism to plan and execute it and developing a comprehensive strategy for it. The administration established the National Security Council (NSC) and its staff, National Security Secretariat, located at the Prime Minister's Office in December 2013. The core of the NSC is the four ministers' meeting consisting of the prime minister, the foreign and defense ministers, and the chief cabinet secretary, who meet regularly.[1] The four ministers' meeting is expected to serve as a control tower for foreign and defense policy related to national security, while larger meetings conduct discussions in a wider context. The first secretary general of the National Security Secretariat, Yachi Shotaro, described this arrangement as follows.

> [The core of] the NSC is a conference of four ministers chaired by the prime minister. The mission of the National Security Secretariat is to provide the conference with materials for discussions and policy decisions as well as policy options. Thus, the secretariat serves as the foundation for a control tower under political leadership for foreign and defense policies.[2]

The Abe administration announced the first NSS in Japanese history on December 17, 2013. The National Defense Program Guidelines (NDPG 2013) for fiscal year 2014 and beyond and the Mid-term Defense Program (MTDP) for fiscal years 2014–2018 were released on the same day based on the NSS, covering a broader context encompassing the defense strategy. A defense strategy and policies to implement it fit into the wider NSS context. In the case of the United States, a national security strategy is issued at the presidential level, which leads to a national defense strategy at the level of the secretary of defense, following which the chairman of the Joint Chiefs of Staff develops a national military strategy. This, in turn, sets the strategic context for subordinate strategies such as those of service chiefs (army, navy, air force, marines) and unified commanders (Pacific Command, etc.). In the past, planners working on NDPGs had to surmise what a national security strategy would describe. While the newly developed NSS may suffer from being just the first of its kind, it provides the defense strategy and policy with a broader context within which defense planners along with those working on diplomatic, economic, and other aspects of national security will be able to comprehend the role of defense in national security as a whole. This change in the process of developing defense strategy and policy will ensure that they are consonant with all other aspects of Japan's security strategy such as those on diplomacy, commerce, and trade, while fitting precisely into a broader picture of NSS. The establishment of the NSC and a permanent staff for it reinforces the planning and execution of strategy in a comprehensive manner.

Proactive Contributor to Peace: Core Concept of the NSS

The key concept of the NSS, Japan as a "proactive contributor to peace," is clearly stated as follows:

> Japan will continue to adhere to the course that it has taken to date as a peace-loving nation, and as a major player in world politics and economy, contribute even more proactively in securing the peace, stability, and prosperity of the international community, while achieving its own security as well as peace and stability in the Asia-Pacific region, as a "Proactive Contributor to Peace" based on the principle of international cooperation. This is the fundamental principle of national security that Japan should stand to.[3]

The NDPG 2013 also states, "Japan will contribute more actively than ever to ensure peace, stability, and prosperity of the world," following "the policy of 'Proactively Contributing to Peace' based on international cooperation."[4] Since 1991, when Japan dispatched the Maritime Self-Defense Force's (MSDF) minesweepers to the Persian Gulf after the Gulf War, and 1992, when Japan for the first time participated in a UN peacekeeping operation (UN PKO) in Cambodia, the SDF has been active in UN PKO, humanitarian assistance/disaster relief missions, and postconflict reconstruction missions. In terms of legal status, however, these missions had been categorized as miscellaneous activities as opposed to primary missions until 2007, when the SDF law was revised to list international cooperation activities as primary missions along with other key missions such as the defense of Japan. The new NDPG declaring that Japan should "proactively contribute to peace" properly placed the international missions of the SDF at the center of national security policy.

In addition to active participation in international peace activities, the administration emphasizes the importance of enhanced cooperation with various partners. The NSS, while mentioning that strengthening the Japan-US alliance is indispensable, states that "Japan will engage itself in building trust and cooperative relations with other partners both within and outside the Asia-Pacific region" to forge a better strategic environment surrounding Japan. In this context, the NSS regards Korea and Australia, as well as other countries with which Japan shares universal values and strategic interests, as important partners next to the United States. It is notable that the NSS recognizes the importance of Sino-Japanese relations stating that "stable relations between

Japan and China are an essential factor for peace and stability of the Asia-Pacific region," and "Japan will strive to construct and enhance a Mutually Beneficial Relationship Based on Common Strategic Interests with China in all areas, including politics, economy, finance, security, culture and personal exchanges."[5] The NDPG 2013 also states that "Japan will promote multi-tiered initiatives, including defense cooperation and exchanges, joint training and exercises, and capacity building assistance."

Dynamic Joint Defense Force: Goal for Defense Buildup

The NDPG 2013 aims at the buildup of a "dynamic joint defense force" with an emphasis on "defense posture buildup in the southwestern region," stating that the priority of the SDF should be placed on capabilities to ensure "maritime and air superiority, which is the prerequisite for effective deterrence and responses in various situations" and capabilities to "deploy and maneuver forces."[6] The NDPG also declares that "the SDF will develop full amphibious capability" in order to deal with an invasion of remote islands.

These basic ideas are further translated into particular programs in the MTDP for fiscal years 2014–2018. Examples of such programs focusing on the defense of remote islands in the southwestern region include a Ground Self-Defense Force (GSDF) watch station and activation of an Air Self-Defense Force (ASDF) early warning squadron equipped with E-2C aircraft as well as GSDF security units in charge of an initial response to contingencies on remote islands in Okinawa prefecture. For maritime and air superiority, procurement of F-35 fighters, Aegis destroyers, and the GSDF's medium range surface-to-air missiles as well as the ASDF's updated Patriot missiles (PAC-3 with Missile Segment Enhancement: MSE) is scheduled. To enhance capability to deploy and maneuver necessary forces, the SDF will introduce Osprey tilt-rotor aircraft while continuing procurement of C-2 cargo aircraft and CH-47J heavy lift helicopters. As to its amphibious capability, the GSDF will procure amphibious vehicles while the MSDF will conduct studies of a new type of ship that supports amphibious operations with command and control, sealift, and aircraft launching capabilities.

In the case of island defense, "joint" is a key word in the new NDPG, meaning cooperation between the ground, maritime, and air SDF. Island defense—amphibious operations in particular—requires an extremely high level of joint operations, where the priority should be given to capabilities to gain air and maritime superiority that guarantees the freedom of maneuver of defending forces. For this, fighters and surveillance assets of the ASDF along with MSDF vessels such as Aegis destroyers with high anti-air combat capabilities should be employed in a harmonized manner. Such anti-air operations should be reinforced by deployment of air defense assets to remote islands in order to provide cover for key facilities such as airports and seaports as well as surveillance stations. In amphibious operations, the highest degree of coordination is required for units from different services, including landing combat forces and the sealift and airlift capabilities to deploy them, as well as maritime and air assets to provide fire power and logistical support to overcome long distances and the obstacles in a marine environment.

Seamless Response to "Gray-Zone" Events: Operational Concept

The NDPG calls for a "seamless response to various situations including so-called 'gray-zone' situations," reflecting the threat perceptions expressed by both the NSS and the new NDPG. For example, the NSS states, "the Asia-Pacific region has become more prone to so-called 'gray-zone' situations, situations that are neither pure peacetime nor contingencies over territorial sovereignty and interests," and "there is a risk that these 'gray-zone' situations could further develop into grave situations." In response, the NDPG emphasizes the need for unified efforts of central and local governments as well as the private sector, stating that "the entire government with strong political leadership will make appropriate and quick decisions, and seamlessly respond to situations as they unfold, in cooperation with local governments and the private sector, in order to ensure the protection of the lives and property of the Japanese people, and the integrity of Japan's territorial land, waters, and airspace."

The response to a crisis, indeed, needs to be seamless in two respects: between different organizations with particular areas of responsibility, and between different rungs of escalation or different phases of events from full peacetime through crisis to serious contingencies. The new NDPG recognizes that the SDF is increasingly required to cope with various situations including "gray-zone" situations, citing the urgency of defining the roles the SDF should and could assume in the effort of the entire government to prevent such "gray-zone" events from escalating to grave incidents such as armed conflicts as well as lowering the tensions toward full peacetime. In this context, the government of

Japan has to work on the legal basis for SDF operations, particularly in the case of "gray-zone" events, as discussed in relation to reconstructing the legal basis for security.

Reconstruction of the Legal Basis for Security

Abe has consistently pressed for the reconstruction of the legal and constitutional basis for Japan's security policies. This is reflected in discussions of the Advisory Panel on Reconstruction of the Legal Basis for Security organized by the first Abe administration and reactivated when Abe took office in 2012. Its report on June 24, 2008, pointed out legal problems associated with four particular eventualities and proposed steps to fix such problems. These four categories include: (1) defense of US naval vessels on the high seas: (2) interception of a ballistic missile that might be on its way to the United States; (3) use of weapons in international peace operations; and (4) logistics support for the operations of other countries participating in the same UN PKO and other activities. The advisory group has been active since February 2013 and is expected to release proposals on the legal basis for security in a broader context, including defense operations to exercise the right of individual self-defense and exercise of the right of collective self-defense as well as problems related to the four categories identified in the previous report.

Japan "interprets Article 9 of the Constitution to mean that armed force can be used to exercise the right of self-defense only when the following three conditions are met: (1) when there is an imminent and illegitimate act of aggression against Japan; (2) when there is no appropriate means to deal with such aggression other than by resorting to the right of self-defense; and (3) when the use of armed force is confined to the minimum necessary level."[7] While this position appears to be similar to the standard based on international law, Japan has been extremely restrictive in exercising the right of self-defense, mainly because it had caused devastation by war in Asia before 1945.[8] In particular, its position on the exercise of collective defense has been restrictive:

> International law permits a state to have the right of collective self-defense, which is the right to use force to stop an armed attack on a foreign country with which the state has close relations, even if the state itself is not under direct attack. Since Japan is a sovereign state, it naturally has the right of collective self-defense under international law. Nevertheless, the Japanese Government believes that the exercise of the

right of collective self-defense exceeds the minimum necessary level of self-defense authorized under Article 9 of the Constitution and is not permissible.[9]

In a Cold War context, this position was relevant to Japan's restrictive defense policy aiming at a peaceful nation since Japan tried its best not to get involved in any possible armed conflict between West and East not directly related to Japan's own self-defense. In the post–Cold War environment, however, it seems necessary to give serious consideration to whether this restraint continues to contribute to Japan's pursuit of peace in the international community.

When Japan moves in the direction dictated by the new NSS as a "proactive contributor to peace," deliberations on constitutional restraints become more important. As the Cold War East-West confrontation disappeared, a cooperative approach toward international security became much more feasible. International peace activities, as UN PKO, have become more frequent—more common than traditional peacekeeping following the termination of armed conflict that includes "peacebuilding" operations. Since 1992 when Japan sent its first peacekeepers to Cambodia, it has actively participated in international peace operations. The JSDF contingents in Cambodia in 1992 worked together with PLA soldiers and with Korean peacekeepers in Timor Leste in 2002–2004 for reconstruction of the two countries. Currently some 400 Japanese are working for the UN Mission in the Republic of South Sudan (UNMISS) with peacekeepers from various militaries, including the Chinese PLA and the Korean armed forces.[10]

The "Report of the Advisory Panel on Reconstruction of the Legal Basis for Security" made a set of proposals on legal problems associated with such international peace operations. It argues that it is "necessary to allow the SDF to come to the aid of a geographically distant unit or personnel of other countries that are engaged in the same UN PKO and other related activities, and to use weapons, if necessary, to defend them, in the event that such a unit or personnel are attacked."[11] This kind of activity by the SDF, involving the use of weapons, has been widely regarded as an exercise of the right of collective defense; thus it could be regarded as exceeding the limits dictated by the constitutional interpretations. During the PKO in Cambodia in 1992–1993, in which Chinese and Japanese peacekeepers worked together, the Chinese contingents suffered from a mortar attack resulting in two fatalities and some dozen wounded. In the South Sudan there have been fatalities, including Indian peacekeepers, while Japanese have been safe. While

UN PKOs are not intended for combat, they are not conducted under perfectly safe conditions. There may be cases where the Japanese contingent needs the assistance of the armed forces of other countries as well as where it is needed to assist units from other countries. In the worst-case scenario, the Japanese contingent would be asked to protect other peacekeepers and not be able to do so due to constitutional restraints. If this meant intentional failure to save fellow peacekeepers from Asian countries such as Korea and China, it could result in another history issue lasting for a number of decades ahead. Restraining from exercising the right of collective defense is obviously not sufficient to show Japan's determination to be a peace-loving nation.

There also exist problems in exercising the right of individual self-defense. According to international law and the Constitution, Japan can use force based on the right of self-defense when three conditions listed earlier are met. As the legal basis for such use of force, Article 88 of the SDF Law states, "SDF personnel and units under defense operations may take necessary military action to defend Japan." The Report of the Advisory Panel on Reconstruction of the Legal Basis for Security notes that "an order to the SDF to conduct defense operations is a prerequisite for Japan to exercise the right of self-defense," and "the GOJ has to follow extremely strict procedures in order to issue that order," which "include deliberation by the Security Council of Japan, followed by a Cabinet Decision, and then prior approval by the Diet." The report warns, "Japan would not be able to effectively respond to new types of threats, such as ballistic missiles and terrorism." It is urgent to remove legal restraints hindering a prompt response to various contingencies, such as small-scale but abrupt armed attacks, and to take necessary measures to promote readiness to cope with them. This problem becomes more serious when considering "gray-zone" events because it is hard to conceive of a timely defense order for use in such cases. In order to prevent and control escalation of such events at the lowest level, it is urgent to establish the legal basis for SDF operations under conditions where a defense order has yet to be issued.

The advisory panel warns, "It would be detrimental to the Japan-US alliance, if Japan did not shoot down a ballistic missile that might be on its way to the United States when Japan is capable of doing so."[12] If such action is regarded as unconstitutional because it involves the right of collective defense, this problem is not necessarily limited to the case with the United States. Since SM-3 missiles with which the MSDF's AEGIS ships are equipped are designed to shoot ballistic missiles during their ascent phase and mid-course, their coverage ranges far if the ship is positioned close to the launchers. In 2012 when North Korea twice announced that it would launch a missile toward the south, the SDF deployed AEGIS destroyers in the East China Sea along with PAC3 land-based ballistic missile defense systems deployed on southwestern islands. Japan's AEGIS ships, if positioned close to North Korea, hypothetically might provide ballistic missile defense for a wide area, including the Philippines, Indonesia, and even the southeastern part of China as well as Japan's southwestern islands. Not shooting down a missile on its way to another country would mean that Japan intentionally avoided protecting people in danger from such a missile even though it had the capability to do so.[13]

Consideration of China's Rise and the US Rebalance toward Asia

The security policies of the Abe administration stated earlier have been declared in obvious awareness of two important trends in Japan's security environment, that is, the US rebalance and China's rise. China's rise is a given, with the only questions remaining being in what direction and how fast. The best scenario for Japan is a strong US commitment to the region and the benign rise of China. Thus, its policies should strive to keep the United States committed and to establish constructive relations with China through engagement policies while hedging to prevent China from going in the wrong direction through its own efforts and its cooperation with the United States.

The NSS recognizes that "the US remains the country that has the world's largest power as a whole, composed of its soft power originating from its values and culture, on top of its military and economic power," while acknowledging changes in the relative influence of the United States in the international community. As to the "rebalance," the NDPG 2013 points out that "the United States has clearly communicated its strategy to put greater emphasis on the Asia-Pacific region and is maintaining and strengthening its involvement and presence in the region despite fiscal and various other constraints, while enhancing its relationship with its allies and other countries."

Japan's assessment of China is cautious or even alarming, although the NSS expresses the expectation that China will "play a more active and cooperative role in regional and global issues." The NDPG 2013 registers concern that China "is rapidly expanding and intensifying its activities in waters and airspace in areas including the South China Sea and the East China Sea, showing its attempts to change the status quo by coercion." Yet, the importance of better relations with China is often noted. The NSS states that "stable relations between Japan and China are an essential factor for the peace and stability of the Asia-Pacific region," and "Japan will strive to construct and enhance a mutually beneficial relationship based on common interests with China" in various areas.

Japan's choices are: to strengthen the alliance with the United States in order to assure its commitment to the region and to build constructive relations with China through engagement while hedging to avoid a situation where Japan has to consider China as a hostile entity. For the alliance, it is important to revise the "Guidelines for Japan-US Defense Cooperation." The first guidelines, adopted in 1978, described operational cooperation between US forces and the SDF and were revised in 1997 to adapt the alliance to the post–Cold War environment. The ongoing efforts to revise the guidelines should be extended to include the bilateral response to "gray-zone" situations as well as bilateral cooperation for other peacetime activities such as counterpiracy and humanitarian assistance/disaster relief operations and to deal with issues related to new domains such as the open seas, outer space, and cyberspace.

As to China's rise, the SDF's shift toward the southwestern region and buildup through a Joint Dynamic Defense Force will work well for hedging purposes. Such policies will also work well to enhance the credibility of the Japan-US alliance by showing Japan's determination to defend itself. Japan also needs to develop and elaborate its strategy and policy for simultaneous efforts at engagement. The NDPG 2013 points out that as China's attitude has great influence over security in the region, Japan, for mutual understanding, "will promote security dialogues and exchanges with China and will develop confidence-building measures to avert or prevent unexpected situations." Since China and Japan are in a state of high tension centering on the sovereignty of the Senkaku (Diaoyu) islands, it is hard for the SDF and PLA to make progress in military-to-military dialogue and exchange programs. Given this level of tension, it is even more important for the two militaries to talk to each other for confidence building.

Conclusion

As pointed out in the *Defense of Japan 2013*, Chinese defense expenditures have been growing around 10 percent per year, and "the nominal size of China's announced national defense budget has approximately quadrupled in size over the past ten years, and has grown more than

33-fold over the past 25 years."[14] Japan's defense budget for the past 20 years has remained at roughly the same level or slightly decreased. Under such circumstances, the SDF has kept in relatively good shape as it has actively participated in international peace activities abroad. In fact, Japan's defense budget from the 1970s until after the collapse of the bubble economy in the mid-1990s had kept growing at a 5 percent annual increase despite the fact that most of the major nations in the world sharply decreased their military spending when the Cold War ended in the late 1980s. While this past build-up helped the SDF manage in the leaner years that followed, such reserves are now almost exhausted. The Abe administration's decision to increase the defense budget by 2.8 percent (including funds for realignment of the US forces in Japan) for the FY 2014 is overdue and appropriate. The SDF must try best to implement every policy for achieving the goal of building up a "Dynamic Joint Defense Force."

Japan's own defense buildup is also important in the context of the Japan-US alliance because it shows the determination to take responsibility as an ally. In parallel to this effort, Japan must be keen about taking measures to reconstruct the legal basis for security that will strengthen the alliance, such as those for the protection of US naval vessels on the open seas and the interception of ballistic missiles that might be on their way to the United States. The two governments are currently working on the new "Guidelines for Japan-US Defense Cooperation," which will provide the two with a golden opportunity to coordinate their respective security policies and to share threat perceptions and security priorities.

Aware of the importance of building and maintaining good relations with Japan's neighbors, the Abe administration has been active diplomatically and successful in improving ties with many countries, notably the members of ASEAN. It remains an urgent task to rebuild constructive relations with Korea and China. In the case of relations between Japan and Korea, both have sometimes become too emotional, giving impetus to voices considered to be anti-Japanese or anti-Korean. The NSS appropriately states, "The ROK is a neighboring country of the utmost geopolitical importance for the security of Japan. For this reason, Japan will construct future-oriented and multilayered relations and strengthen the foundation for security cooperation with the ROK. In particular, trilateral cooperation among Japan, the US, and the ROK is a key framework in realizing peace and stability in East Asia. Japan will strengthen this trilateral framework, including cooperation on North Korean nuclear and missile issues." The Japanese and

Koreans should share a common understanding of the importance of the Korea-US and Japan-US alliances, in the sense that the former provides Japan with the security of the Korean Peninsula and the latter lets Korea enjoy the security of its backyard in case of Korean contingencies. Sino-Japanese relations have been at an extremely high level of tension since September 2012 when the government of Japan decided to purchase the Senkaku islands. Chinese and Japanese maritime law enforcement organizations are currently in an eyeball-to-eyeball confrontation on a daily basis. It is urgent to establish confidence-building measures between the two countries to avoid unnecessary and unintentional escalation after an accident, such as a collision of ships or aircraft, as part of further efforts of the two countries to build constructive relations, as stated in the NSS.[15]

Notes

1. Ministry of Defense, *Defense of Japan 2013*, 105–106, http://www.mod.go .jp/e/publ/w_paper/index.html (accessed March 30, 2014).
2. Suzuki Yoshikatsu, "Yachi shodai Kokka Anzen Hosho kyokucho ni kiku," *Gaiko* 23 (January 2014): 68.
3. National Security Strategy (NSS), released by the government of Japan on December 17, 2014 (provisional translation), http://www.cas.go.jp/jp/siryou /131217anzenhoshou/nss-e.pdf (accessed March 30, 2014).
4. National Defense Program Guidelines for FY 2014 and beyond (NDPG 2013), http://www.mod.go.jp/j/approach/agenda/guideline/2014/pdf/20131217 _e2.pdf.
5. NSS.
6. NDPG 2013.
7. Ministry of Defense, *Defense of Japan 2013*, 101.
8. Taoka Ryoichi, *Kokusaiho-jo no jieiken* (Tokyo: Keiso-shobo, 1985), 32–47. It has been understood by the international community that use of armed forces based on the right of self-defense should be limited to cases where the necessity of self-defense was instant, overwhelming, leaving no choice of means, and no moment of deliberation, and such use of force must not be unreasonable or excessive, and must be limited by that necessity, and kept clearly within it.
9. Ministry of Defense, *Defense of Japan 2013*, 101.
10. Ministry of Defense, "About the JSDF Activities in South Sudan (February 12, 2014)," http://www.mod.go.jp/j/approach/kokusai_heiwa/s_sudan_pko/pdf /gaiyou.pdf (accessed March 30, 2014).
11. "Report of the Advisory Panel on Reconstruction of the Legal Basis for Security," http://www.kantei.go.jp/jp/singi/anzenhosyou/report.pdf (accessed March 30, 2014).
12. "Report of the Advisory Panel on Reconstruction of the Legal Basis for Security."

13. For further discussion on the constitutional debate, see Noboru Yamaguchi, "*Shudanteki jieiken o giron suru maeni kangaerubeki koto*," posted by *Diamond Online*, http://diamond.jp/articles/-/47195 (accessed March 30, 2014).
14. Ministry of Defense, *Defense of Japan 2013*, 33.
15. The author's ideas on the management of Sino-Japanese tensions are expressed further in an article posted by the Lowy Institute, Yamaguchi Noboru, "A Japanese Perspective on the Senkaku/Diaoyu Islands Crisis," http://www.lowyinstitute.org/publications/tensions-east-china-sea (accessed March 30, 2014).

CHAPTER 10

Japanese Politics Concerning Collective Self-Defense

Yuichi Hosoya

Part 1

When Abe Shinzo entered a meeting room in the Prime Minister's Office in the afternoon of May 15, 2014, a group of Japanese journalists awaited his arrival to take a picture of an historic moment. Members of the Advisory Panel on Reconstruction of the Legal Basis for Security, including myself, were also there and due to submit our final report to him recommending that the government revise the previous constitutional interpretation on the exercise of the right of collective self-defense. The submission of this report, as expected, aroused a great political debate in Japan, which has not abated.

Chairman Yanai Shunji, who was previously administrative vice minister of foreign affairs, handed the report to Abe. Seven years had passed since Abe had established the advisory panel in April 2007. His original plan to revise the constitutional interpretation on the exercise of collective self-defense had failed, when Abe stepped down as prime minister in September 2008 due to illness. The next prime minister, Fukuda Yasuo, was reluctant to take up this issue; his priority was to enhance Japan's friendly relationship with China. There existed many barriers that prevented revising the constitutional interpretation, which was formulated by the Cabinet Legislation Bureau in 1981. Yet, leadership mattered, and when Abe returned to this office in December 2012, he showed his strong will to reconvene the panel, which happened in

February 2013. I became a member in September 2013 and then collaborated in drafting the report.

When Abe received the report of the advisory panel, he issued a short statement thanking the group for its submission and expressing gratitude "for the intensive discussion and many valuable opinions we have received over the years from everyone on the panel regarding the ideal form of the legal basis for security."[1] Abe looked both satisfied and relaxed as he had finally reached an important goal.

Abe's ambition to revise the constitutional interpretation concerning the right of collective self-defense has aroused fierce political controversy. Japanese newspapers have covered this issue nearly every day. Protesters who are against lifting the ban on the exercise of collective self-defense appear around the Prime Minister's Office on a daily basis. Abe and his chief cabinet secretary Suga Yoshihide are being pressed to comment on this issue in public and are obliged to clear doubts about the necessity of the reinterpretation. This issue divides both Japanese politicians and the media. Conservative *Yomiuri Shimbun* and right-wing *Sankei Shimbun*, together with *Nikkei Shimbun*, have clearly been supporting Abe's move to revise the interpretation, but progressive *Asahi Shimbun* and *Mainichi Shimbun* are severely criticizing this shift.

The LDP is basically united in supporting its own prime minister on this issue, with a few exceptions. The New Komeito, its coalition partner, resisted the radical change in constitutional reinterpretation. The largest opposition party, the DPJ, has been divided, and it cannot decide what its party line should be. The issue of collective self-defense has kicked up a political storm, which is occupying much of the political space in Japan.

This issue has become extremely complex. Besides, the coalition politics between the LDP and the New Komeito have further complicated matters. There are many compromises and concessions on both sides. It is important to clearly understand what the report of the advisory panel recommended and also what the cabinet decided. As a member of the panel, I can present what the advisory panel really intended in the report.

The Role of the Advisory Panel

Following a respite, the government decided to restart the working of the advisory panel after the LDP won the Upper House elections in July 2013. It convened six times from September 2013 to May 2014 at the Prime Minister's Office. Abe attended every meeting and actively joined in the discussion; he regarded the issue as one of his most important policy agendas. He wrote in his book, "The future of Japan will be decided by whether the government can revise the constitutional interpretation of Article 9 and can exercise the right of collective self-defense."[2] As the strategic environment in East Asia has deteriorated over the past seven years, due largely to the rapid expansion of Chinese military activities, Abe asked the advisory panel to write a report that would reexamine Japan's legal basis for security. This was the objective that guided its deliberations.

The basic role of the advisory panel was to discuss a variety of issues concerning the legal basis for security. The issue of the constitutional reinterpretation of collective self-defense was one. People mistakenly thought that the advisory panel was only for the purpose of revising this interpretation. The most imminent issues included legal arrangements for "gray zone" events and the issue of the use of weapons in UN peace-keeping operations. Abe often repeated the phrase, "proactive contribution to peace based on international cooperation (*kokusai kyochoshugi ni motozuku sekkyokuteki heiwashugi*)," suggesting a scope for the advisory panel far broader than people assumed. Both the government and the advisory panel thought that the legal basis for security in Japan has to accommodate the recent changes in Japan's security environment. It is difficult for ordinary people to grasp the overall picture of what the report of the advisory panel intended to recommend. It is often the case that people tend to be influenced by Japanese media coverage. Japanese newspapers are ideologically polarized, and the two sides are sharply at odds; reading the media can be misleading.

Public Opinion in Flux

It is not easy to say whether the public in Japan is resisting the constitutional reinterpretation or not, as the results of opinion polls largely differ depending on newspaper companies. According to the opinion poll on April 19–20 by *Asahi Shimbun*, which heavily criticized lifting the ban on the exercise of the right of collective self-defense, it was reported that only 27 percent of respondents supported the constitutional reinterpretation, while 56 percent were against it. In contrast, *Yomiuri Shimbun* reported in its own opinion poll of May 9–11 that 72 percent supported the revision of the constitutional interpretation, while only 25 percent did not.[3] This gap suggests that the respondents answered differently according to the wording of the question.[4] Japanese opinion polls are conducted mainly by major newspaper companies, each of which maneuvers to exaggerate its desired result. The truth is that a majority of people do not fully understand this difficult legal and security issue, even if it appears that many of the Japanese people are inclined to stick to the traditional pacifist norm.

While the recommendations of the advisory panel are both comprehensive and complex, *Asahi Shimbun* simplified the argument by saying that "Prime Minister Abe radically destroys Japan's pacifism, which had been the central pillar of the Constitution of Japan."[5] Those who repeatedly read such criticisms would naturally feel anxiety about Abe's seemingly "dangerous" move toward war. On the other side, Japanese security experts are arguing that this will broaden Japan's proactive contribution to international peace and stability. Confusion results from the difficulty of clearly understanding what the report of the advisory panel actually has recommended to broaden this contribution.

The Report and the Cabinet Decision

In the introduction to the report of the panel, it is written that "the security environment surrounding Japan has changed dramatically even in the few years since the Panel submitted its previous report."[6] Then, "this has necessitated serious consideration in Japan's security policy towards the maintenance and building of peace in the international community." The report clearly argues that the legal basis for security has to be adapted to the changing security environment surrounding Japan. Thus, the previous constitutional interpretation of collective self-defense ought to be revised.

In the second section, the advisory panel clarifies "how the Constitution should be interpreted,"[7] expecting that the government would effectively make a Cabinet Decision, and then the National Diet would need to legislate a revised Self-Defense Forces (SDF) law. Based on these new legal arrangements, the SDF would be able to act in areas that were not previously permitted. The report includes "participation in collective security measures of the United Nations entailing military measures." Other areas mentioned are: "cooperation and the use of weapons in UN PKOs, etc." and "response to an infringement that does not amount to an armed attack." These areas are not directly linked to the right of collective self-defense. The former relates to the issue of PKO operations and the latter to individual self-defense. The panel recommends revising the SDF Law in many ways, as the current security environment necessitates.

The New Komeito does not want to adopt all of the recommendations of the advisory panel. Its president, Yamaguchi Natsuo, has been reluctant to revise the previous constitutional interpretation, an action his party regards as the abandonment of Japan's postwar peaceful path. The Buddhist organization Soka Gakkai, which is the religious body of the New Komeito, has had a leftist political ideology critical of Japan's broader contribution to international peace and security. After the submission of the report, Yamaguchi has been saying that the government should not permit the exercise of collective self-defense, as this would lead the SDF to become involved in foreign wars.

During May and June 2014, the LDP had been negotiating with the New Komeito on how the government would formulate the Cabinet Decision. New Komeito deputy president Kitagawa Kazuo was responsible for these negotiations, strongly demanding to limit the exercise of collective self-defense to cases when "the situation should pose a clear threat to the Japanese state or could fundamentally threaten the Japanese peoples' constitutional right to life, liberty, and the pursuit for happiness."[8] The New Komeito collaborated with the Cabinet Legislation Bureau for the purpose of finding wording for the new constitutional interpretation. In this way, it thought that postwar pacifism was preserved. For New Komeito, postwar pacifism essentially signifies the limitation of Japanese defense policy to the right of individual self-defense.

On the exercise of self-defense, the government presented three new conditions: (1) the situation should pose a clear threat to the Japanese state or could fundamentally threaten the Japanese peoples' constitutional right to life, liberty, and the pursuit of happiness; (2) there is no other way to repel the attack and protect Japan and its people; and (3) the use of force is limited to the minimum amount necessary. If all three conditions are met, the government is allowed to exercise the right of collective self-defense. Without doubt, the New Komeito will further limit the possibility of exercising this right.

The Essence of the Cabinet Decision

The Cabinet Decision is largely the result of New Komeito's adherence to limiting the decision to individual collective self-defense. The Cabinet Decision rules out participation in both operations of collective security except for minesweeping operations and foreign wars.

On July 1, 2014, the government publicized the "Cabinet Decision on Development of Seamless Security Legislation to Ensure Japan's Survival and Protect its People."[9] A big difference between the report and the Cabinet Decision, due largely to the insistence of New Komeito, is that the latter rules out Japan's participation in collective security operations. Japan will not be able to dispatch the SDF to foreign wars such as the Gulf War of 1991, the Afghanistan War of 2001, and the Iraq War of 2003. Japan's role in this sort of war will be largely limited, at most, to logistical support.

The Cabinet Decision mentions three security activities in which the SDF will be able to operate with the passing of new legislation. The first section is on the "response to an infringement that does not amount to an armed attack." This is the issue of individual self-defense. So far, the SDF can only be mobilized when an organized and planned armed attack occurs. To defend Japan's remote islands, it would be necessary to effectively respond to "an infringement that does not amount to an armed attack."

The second section is on "further contributions to the peace and stability of the international community." By passing new legislation, the Japanese government will be able to further contribute to the international community in the area of logistical support. Since 1997, the Cabinet Legislation Bureau has largely banned so-called logistics support to international coalition forces by creating a concept of "*ittaika* with the use of force." By revising the previous constitutional interpretation, Japan can broaden its contribution in the area of logistical support. In this section, it is written that the limits on the use of weapons associated with international peacekeeping operation activities should be loosened. Otherwise, it would be difficult for the SDF to effectively operate in areas where risks and instability are present.

The third section is the most important one and relates to "measures for self-defense permitted under Article 9 of the Constitution." In this section, the cabinet clearly states that the previous constitutional interpretation should be revised, for the purpose of enabling the exercise of the right of collective self-defense under limited circumstances. "The Government has reached a conclusion that not only when an armed attack against Japan occurs but also when an armed attack against a foreign country that is in a close relationship with Japan occurs and as a result threatens Japan's survival and poses a clear danger to fundamentally overturn the people's right to life, liberty, and the pursuit of happiness, and when there is no other appropriate means available to repel the attack and ensure Japan's survival and protect its people, use of force, to the minimum extent necessary, should be interpreted to be permitted under the Constitution as measures for self-defense in accordance with the basic logic of the Government's view to date."

Long Live *Senshu-boei*!

It is obvious, as seen in the document of the Cabinet Decision, that the government has revised the previous constitutional interpretation to a very limited extent. Japanese newspapers such as *Asahi Shimbun* exceedingly criticized this Cabinet Decision, arguing that this would be a drastic change in Japanese security policy. However, security experts largely agree on the fact that this reinterpretation is both quite limited and incremental. Michael Green and Jeffrey Hornung appropriately said that "Japan's policy will still be based on the decades-old metric of minimal force necessary for exclusively defensive defense *(senshu-boei)*."[10] James Schoff says that "the reinterpretation was not quite as radical as some Americans had hoped for, but will pave the way for the two armed forces to plan, train and operate more seamlessly."[11]

With this reinterpretation, the US-Japan alliance can be further strengthened. The two governments are now working on revising the defense guidelines for the first time in 17 years. Japan can now broaden logistical support to the United States or other friendly forces. As this series of articles continues, we will follow the Japanese further and reflect on views about how the changes in interpretation will affect the US-Japan alliance and the prospects for peace and stability in East Asia.

Rejoinder

Gilbert Rozman

Abe's success in securing cabinet approval—a matter much more complicated in Japan than elsewhere—for a statement on revising the constitutional interpretation of the right of collective self-defense is ably explained by Hosoya. He also discusses the fierce opposition to this in the Japanese media and the importance of central newspapers in shaping public opinion. International observers find it difficult to grasp what is driving the negative response to the advisory panel's report, the Cabinet Decision, and Abe's push for new legislation. What does it mean to stick with Japan's postwar peaceful or pacifist path in the context of 2014? What does it mean to rule out collective security operations in East Asia as opposed to in Afghanistan and Iraq? Is anything beyond logistical support possible? Or is the third section that gives approval for collective self-defense a justification for much more support, such as in an attack on South Korea or Vietnam? What are the implications for revisions of the defense guidelines with the United States expected in 2015? These are questions that readers may raise from reading Hosoya's first statement.

Reading the Japanese press in 2014 raises additional questions about the debate in progress. Are the newspapers stuck in fixed positions or has the debate continued to evolve to late August? How do critics of the Cabinet Decision propose to meet recent security challenges facing Japan, whether from China or North Korea? What foreign policy attitudes—toward South Korea relations, the abductions talks with North Korea, defiance of the West in diplomacy with Russia, priority for Indian relations, a new security relationship with Australia, and so on—correlate with opposition to the new interpretation of collective self-defense? While the Japanese government's position is often presented in Washington, DC and other capitals, the opposition's thinking is a mystery to many. Is it based on a strongly held view of Japanese national identity that many have underestimated as they focused on Nihonjinron in the 1980s and revisionism in the past decade? Is it linked to a different conception of Asianism able to survive the debacle of the DPJ's "fraternal" overtures to China? Or is it based on a different definition of national interests, which sees an opportunity for Japan to be a bridge or to join an East Asian economic community? Do opponents also stand against TPP? There are lots of questions about the divisions inside Japan that have been brought into the open by the collective self-defense issue at a time of general support for the US alliance, alarm about the China threat, and quest for a new Japan.

Part 2

As I wrote in the opening statement, the decision by the Japanese government to revise the previous constitutional interpretation on the right of collective self-defense aroused a huge political debate in Japan. As Gilbert Rozman wrote in his rejoinder, "The opposition's thinking is a mystery to many."

Although the opposition gathered around the Prime Minister's Office nearly every day in June and July, ordinary people did not show strong interest in this complex and difficult legal issue. By the time when Prime Minister Abe reshuffled his cabinet on September 3, 2014, the support rate for the Abe cabinet had increased by 13 percent, reaching 64 percent, according to a poll by *Yomiuri Shimbun*.[12] It is very rare for a Japanese prime minister, except for Koizumi Junichiro, to keep such a high support rate for 20 months. This tells us that the Cabinet Decision on the revision of the constitutional interpretation on collective self-defense has not seriously damaged his popularity.

Those who are against this Cabinet Decision are basically also against TPP and the strengthening of the US-Japan alliance. I regard this as a division inside Japan between "internationalists" and "isolationists" on security policy. What Abe is trying to do is to internationalize Japan's security policy, as his basic doctrine is labeled "proactive contribution to peace based on international cooperation." For this purpose, he considers it necessary to revise the previous, quite isolationist constitutional interpretation. For the opposition group, the most important issue is whether Japan will be entangled in foreign wars. The Japanese people, they argue, should be detached from these foreign wars. They regard this sort of insular pacifism as the most precious value stipulated by Article 9 of the Japanese Constitution. It is widely believed in Japan by those left-wing pacifists that wars can be abolished by simply spreading Article 9 of the Japanese Constitution to the world. This way of thinking is often seen as "a mystery to many."

Having seen strong opposition to the revision of the previous constitutional interpretation, we can now notice two important features of Japanese society concerning security issues. First, the tradition of pacifism is much stronger than people usually assumed. Japanese society exceedingly hates wars and Japan's involvement with foreign wars. Pacifism remains the most respected value for many Japanese people. This results in strong antagonism to broadening Japan's contribution to international peace and stability. This way of thinking is largely based on the assumption that Japan will never be invaded since it has a pacifist constitution.

Second, security policy has been formulated by a small group of security experts in Japan, and ordinary people have usually been detached from security issues. In major Japanese universities, there are no courses on security studies, and military affairs are foreign to a majority of Japanese people. Japanese pacifism is not based on rational thinking on security challenges, but is often based upon irrational ignorance of these challenges.

Abe and Chief Cabinet Secretary Suga Yoshihide are now fully aware of the importance of explaining security policy issues more attentively to the general public. The Japanese government also has to persuade neighboring countries that Japan's larger contribution to international peace and stability can be beneficial to them. This is facilitated by the fact that the Japanese government is still sticking to key features of pacifism, and it is aspiring to bring peace and stability in the Asia-Pacific. The central question is whether the government will be successful in convincing both domestic and international public opinion that Japan's larger contribution can reduce the possibility of a war.

Japanese leaders, including Abe, have repeatedly made this argument, and the public opinion data available leave no doubt that the population at large has no intention to project power in any way that would be threatening to other states. Yet, for domestic and international audiences there remains a challenge to make the case more persuasive. Among various possibilities, four are likely to garner support from one audience or another. First, there are many, including some internationalists in Japan and far more in the international community, who think that the message of unabashed antipathy to militarism and unequivocal support for peace and stability would be strengthened by avoidance of historical revisionism, especially at this sensitive time. Second, the academic and think tank community in Japan needs to rally behind this positive message, introducing courses on security and flooding the media with supportive messages, rather than participating in the campaign by some on the right to concentrate on demonizing progressives as they at last anticipate settling scores dating to the Cold War period. Third, while there is little prospect that China would drop its attacks on the revival of "militarism" in Japan, more should be done to try to reduce South Korean skepticism about the new interpretation of collective self-defense, which, after all, is in the interest of that country's security. Finally, persuading the Japanese and South Korean public is best done through an international effort, including Americans, Australians, and others concentrating on the limited, stabilizing objectives of joint defense efforts.

The fact that the progressive media blitz and demonstrations against the new constitutional interpretation have failed to blunt Abe's popularity is a positive sign that the Japanese public is changing its way of thinking. This is not, however, a reason to be complacent. All of the hot-button issues stirring isolationists are likely to arise simultaneously as TPP, new defense guidelines with the United States, and legislation in support of collective self-defense are debated in 2015. There will be no better time to make the overall case for far-reaching Japanese internationalism.

Rejoinder

Mike Mochizuki

Commenting on Hosoya Yuichi's analysis of the politics of collective self-defense in Japan, Gilbert Rozman noted that international observers are having difficulty understanding what is behind the negative response in the Japanese public to the Abe government's effort to change the legal basis for security policy. In his reply, Hosoya explained this negative response by highlighting two factors. First, he rightly acknowledges that Japan's postwar tradition of pacifism has been remarkably resilient. Second, Hosoya asserts that because ordinary Japanese people have been "detached from security issues" and do not have the opportunity to study security and military affairs in universities, Japanese pacifism "is not based on rational thinking on security challenges, but is often based upon irrational ignorance of these challenges."

In my view, Hosoya dismisses too quickly this pacifism as simply a reflection of "irrational ignorance." He ignores the rich tradition of postwar peace and conflict studies in Japan that takes a skeptical view of the utility of military force in addressing international conflicts. Indeed one does not have to be a hard-core pacifist to recognize how the use of force in a number of cases exacerbated security problems rather than solved them. There are a number of influential Western scholars who are knowledgeable and rational realists who have been quite critical of US military interventions.

I completely agree with Hosoya that "the central question is whether the government will be successful in convincing both domestic and international public opinion that Japan's larger contribution can reduce the possibility of a war." But this requires discussing more concretely and thoroughly what is meant by Japan's larger contribution and how this contribution indeed reduces the possibility of war. A facile application of deterrence theory as a simple slogan (as we now often see in Japanese security policy discourse) is not adequate because informed specialists of security understand the limits of deterrence, the risks of adversary security dilemmas, and the importance of prudent and courageous diplomacy to defuse international conflicts.

Some of the examples that the Japanese government has highlighted to justify the exercise of the right of collective self-defense seem quite far-fetched to those who have a rational and informed perspective on security matters. Take, for example, the frequently mentioned case of Japan needing to shoot down a ballistic missile headed to the US mainland. First of all, it is quite doubtful that Japan will have the ability to take on this mission anytime soon. Second, if a state is about to launch a ballistic missile at the US mainland, it seems much better to have the United States concentrate on shooting down that missile and to have Japan concentrate on defending the Japanese homeland and US military bases in Japan during such a high-intensity conflict scenario. This latter mission can be performed by Japan as an exercise of its right of individual self-defense.

Another unconvincing case that the Abe government has mentioned to justify collective self-defense is the mission of protecting US ships evacuating Japanese nationals from the Korean Peninsula during a military conflict. Isn't it more rational to have Japan take primary responsibility for the evacuation of Japanese nationals rather than hypothesizing that the United States would undertake this mission with the protection of Japanese Maritime Self-Defense Force ships?

Another aspect of the Abe government's handling of the security issue that I find wanting relates to its relations with the Republic of Korea. Insofar as many of the scenarios in which Japan might have to consider exercising the right of collective self-defense concerns the Korean Peninsula, I find it perplexing that Prime Minister Abe has not done more to promote reconciliation with South Korea regarding historical issues and to get South Korea's understanding and support for Japan's reinterpretation of the Constitution. Should this not also be part of Japan's proactive contribution to peace?

Despite these misgivings, I want to stress that I have been for quite some time a vigorous advocate of Japan playing a greater role in UN collective security missions and being able to exercise the right of collective self-defense. My main reasoning for this is not because I want to see Japan become more like the United States, Britain, or France regarding the use of force overseas. Rather, I have supported Japan's constitutional reinterpretation because I admire its skepticism regarding the use of force and I respect its post–World War II pacifism. Therefore, I would like to see Japan's voice enhanced in the international deliberative process regarding when the use of force may be legitimate and necessary.

Part 3

Mike Michizuki, with his deep understanding of how the Japanese people perceive recent changes in security policy, is right in saying that "a facile application of deterrence theory as a simple slogan (as we now often see in Japanese security policy discourse) is not adequate because informed specialists of security understand the limits of deterrence, the risks of adversary security dilemmas, and the importance of prudent and courageous diplomacy to defuse international conflicts." I agree that the Japanese government should acknowledge "the importance of prudent and courageous diplomacy to defuse international conflicts." However, "prudent and courageous diplomacy" should be adopted by China as well. One of the biggest questions for the future peace and stability in the Asia-Pacific is whether the Chinese government would sufficiently respect "the importance of prudent and courageous diplomacy to defuse international conflicts" both in the South China Sea and the East China Sea. If it has no intention to do so, Japan's efforts will be in vain.

Those who severely criticize Japan's security policy often dismiss China's assertive military actions, which cause uncertainty in neighboring countries. According to an opinion poll of April 2014, 65 percent of the ASEAN people responded that Japan is the more important partner, while 48 percent responded China.[13] This clearly indicates that more people in ASEAN now welcome Abe's approach to peace and security and fewer consider that China is the more important partner, due largely to the increasingly assertive attitude of China. The criticism toward Japan should be proportional and balanced. David A. Welch rightly argues that "Japan is the least militaristic country I know, with the possible exception of Iceland."[14] His comment is consistent with Mochizuki's understanding that "Japan's postwar tradition of pacifism has been remarkably resilient." As Mochizuki rightly argues, it is now necessary to ask "how this contribution indeed reduces the possibility of war."

My basic argument is that, regardless of his often nationalistic posture, Abe intends to maintain Japan's pacifism and endeavors to "reduce the possibility of war." The suspicion that Abe is planning a war to destroy peace and stability in the Asia Pacific would be a "dangerous false fear," as Welch points out. Abe repeatedly argues that international conflicts should be solved by the rule of law. In his interview with *Foreign Affairs*, Abe clearly said, "Throughout my first and current terms as prime minister, I have expressed a number of times the deep remorse that I share for the tremendous damage and suffering Japan caused in the past to the people of many countries, particularly in Asia."[15]

While Japan's resilient pacifism seems highly evaluated by both international and domestic public opinion, it is also often pointed out that Japan has not sufficiently contributed to international peace and stability. Mochizuki argues that he has been "for quite some time a vigorous advocate of Japan playing a greater role in UN collective security missions and being able to exercise the right of collective self-defense." He says that he admires "Japan's skepticism regarding the use of force" and respects "its post–World War II pacifism." Such "skepticism regarding the use of force" can also be seen in the Cabinet Decision.

Due partly to the role of the New Komeito and the Cabinet Legislation Bureau, the Cabinet Decision on "Development of Seamless Security

Legislation to Ensure Japan's Survival and Protect its People" of July 1 reflected "Japan's skepticism regarding the use of force." I wrote above, "The Cabinet Decision is largely the result of New Komeito's adherence to limiting the decision to individual collective self-defense." This is because "the Cabinet Decision rules out participation in both operations of collective security except for minesweeping operations and foreign wars." Therefore, those who expected Japan's broader contribution to international peace and security are largely disappointed at this Cabinet Decision.

It seems that Mochizuki is uncertain about how the constitutional reinterpretation under Abe's Cabinet will affect "Japan's postwar tradition of pacifism." In the Cabinet Decision, it is clearly stated that "in order to adopt to the changes in the security environment surrounding Japan and to fulfill its responsibility, the Government, first and foremost, has to create a stable and predictable international environment and prevent the emergence of threats by advancing vibrant diplomacy with sufficient institutional capabilities, and has to pursue peaceful settlement of disputes by acting in accordance with international law and giving emphasis to the rule of law." Is this not exactly what Mochizuki expects for Japan or is this likely to be the beginning of Japan's dangerous militarism?

Notes

1. Meeting of the Advisory Panel on Reconstruction of the Legal Basis for Security, May 15, 2014, http://japan.kantei.go.jp/96_abe/actions/201405/15anpo.html.
2. Abe Shinzo, *Kiseki: Abe Shinzo goroku* (Tokyo: Kairyusha, 2013), 40.
3. *Asahi Shimbun* reported this gap in opinion polls by different institutions on May 14, 2014.
4. *Asahi Shimbun*, May 14, 2014.
5. *Asahi Shimbun*, July 3, 2014.
6. The Advisory Panel on Reconstruction of the Legal Basis for Security, *Report of the Advisory Panel on Reconstruction of the Legal Basis for Security*, May 15, 2014, 3.
7. *Report of the Advisory Panel on Reconstruction of the Legal Basis for Security*, 22–45.
8. *Yomiuri Shimbun*, July 16, 2014.
9. "Cabinet Decision on Development of Seamless Security Legislation to Ensure Japan's Survival and Protect Its People," July 1, 2014.
10. Michael Green and Jeffrey W. Hornung, "Ten Myths about Japan's Collective Self-Defense Change," *The Diplomat*, July 10, 2014, http://thediplomat.com/2014/07/ten-myths-about-japans-collective-self-defense-change/.

11. *The Economist*, July 5, 2014.
12. *Yomiuri Shimbun*, September 5, 2014.
13. *Yomiuri Shimbun*, April 20, 2014.
14. David A. Welch, "The Dangerous False Fear of Japanese Militarism," *Asahi Shimbun*, October 4, 2014.
15. Abe Shinzo, "Japan Is Back: A Conversation with Shinzo Abe," *Foreign Affairs*, July/August, 2013, 5.

PART IV

Obama's Meetings with Abe and Park in March and April 2014 and US-Japan Relations in Late 2014

CHAPTER 11

Circling the Square: History and Alliance Management

Lee Chung Min

As President Barack Obama prepares to visit Asia in April 2014 with key stops in Seoul and Tokyo (most likely his last visit to these two capitals as president), the 64 dollar question is whether he will be able to foster Korean-Japanese rapprochement. Although a bilateral Korean-Japanese summit is still unlikely in the near term, a trilateral ROK-US-Japan is slated to be held in The Hague between March 24 and 25 on the sidelines of the 2014 Nuclear Security Summit. From a US perspective, the ongoing, mini-cold war (not to mention an even more volatile Sino-Japanese standoff) is highly disconcerting, given that South Korea and Japan are America's closest and most important allies in Asia. Nevertheless (and notwithstanding core common security interests that bind Washington, Seoul, and Tokyo), it is time to put in place a new alliance management paradigm. Seoul recognizes the importance of sustained trilateral cooperation over a range of critical issues, but not at the expense of ignoring or downplaying Japan's historical amnesia and whitewashing of wartime atrocities.

For the past six decades, the ROK-US alliance and the US-Japan alliance have not only served to maintain peace and prosperity in East Asia, they have also come to symbolize two major success stories in post–World War II US foreign policy. Japan is the world's third largest economic power while South Korea is the fourteenth. Both are democracies with growing contributions to the global commons. Japan is one of the biggest donors to the UN. Their security continues to be inextricably linked by virtue of shared strategic interests and, crucially, due to their alliances with the United States. Thus, it may seem baffling, in the extreme, to outside observers that what has worked so well over the past six decades—America's Asian alliances, including the Washington-Seoul-Tokyo partnership—is suddenly in need of urgent repair. Many are asking what the core problem is and how it can be fixed or, at a minimum, ameliorated?

Perceptions cannot but differ depending on contrasting national viewpoints and historical trajectories, but from a South Korean perspective, business as usual with Japan (especially in the security and political sectors) is no longer possible. To begin with, regardless of one's political persuasion and standing, there is wide-ranging consensus in South Korea that Prime Minister Abe Shinzo's more nationalistic policies are being pursued in conjunction with historical revisionism. Over a range of extremely sensitive issues such as compensation for the remaining survivors who were forced into sexual slavery by the Japanese military during World War II, Japanese textbooks that water down or even ignore extremely cruel legacies of Japanese colonialism and aggression such as the Nanjing massacre of 1937 where some 300,000 Chinese and other nationalities were slaughtered by the Imperial Japanese Army, or the horrible tortures and grotesque medical experiments conducted by Unit 731 headed by Dr. Tamiya Takeo—Japan's version of Nazi Germany's Josef Mengele—there is near universal consensus in South Korea that Japan's obliviousness to key historical tragedies can no longer be tolerated.

Those who are strong supporters in the United States of the ROK-US alliance may believe that South Korea is overreacting to statements made by seemingly cavalier Japanese officials. They point to the fact that the Obama administration expressed "disappointment" with Abe's visit to the Yasukuni Shrine in December 2013 despite multiple signals not to do so and that the US government has spoken out strongly against blatant absurdities, such as NHK board member Hyakuta Naoki's statement that the Nanjing massacre never occurred and that the Tokyo Tribunal was staged by the Americans to cover up their own war crimes such as the atomic bombings in Hiroshima and Nagasaki. Although Seoul is mindful of the recent statements by the Obama administration, it is no longer willing to downplay gross historical distortions on the part of the Japanese leadership in order to move forward policy coordination on critical security issues such as North Korea's WMD threat.

Some foreign commentators have noted that South Korea's "harsh reactions" toward the Abe government may stem from a combination of factors such as nascent negative feelings toward Japan that remained bottled up until democratization in 1987. Others maintain that South Korea's responses to Japan's historical revisionism are a partial reflection of a growing tilt toward China as Beijing reassumes center stage in Northeast Asia's geopolitics. There are those who also assert that anti-Japanese sentiment is so deep and universal because it serves to solidify South Korea's self-identity in the midst of the world's largest and most powerful countries. Such interpretations, however, fail to recognize three critical facets of South Korea's ongoing condemnation of Japan's historical amnesia.

First, although South Korea is most sensitive to the brutal legacies of Japan's colonialism from 1910 to 1945 and especially from 1937 when Japan invaded China until its defeat in 1945, Koreans consider Japan's inability to come to terms with its own dark history a reflection of broader historical amnesia, for example, the fact that many Japanese feel that Japan was the victim rather than the aggressor during World War II. For much of the postwar period with the notable exception of the extreme far right, notions of Japanese victimhood were buried by the more urgent task of rebuilding Japan into an economic powerhouse, jointly fighting the Cold War together with the United States, South Korea, and other American allies such as Australia, and the public's rejection of any real moves toward a remilitarized security policy. Today, South Koreans believe that nearly 70 years after the end of World War II, Japan's self-imposed period of atonement is nearing its end.

Second, according to an opinion survey that was conducted by the Asan Institute for Policy Studies in January 2014, some 76.5 percent of South Koreans believe that Japan shifted to the right under Abe's leadership, but more importantly, South Korea's favorability ratings of Japan have fallen to an all-time low. In the same survey, the favorability ratings of key neighbors on a 1–7 scale (1 being most negative) were as follows: the United States (5.5), China (4.6), Japan (2.4), and North Korea (2.1). Perceptions of Abe at 1.0 (just below the level for Kim Jong-un) compared to Barack Obama (6.2), Xi Jinping (4.6), and Vladimir Putin (4.1). Over time, public perceptions about Japan could improve, but so long as South Koreans continue to perceive that Japan's rightward revisionism is unlikely to budge, these extremely negative views about its political leadership and the country at large seem destined to continue.

Third, it is critical to keep in mind that South Korea's deeply rooted feelings against Japan's historical amnesia remain very separate from other key anxieties such as the longer-term ramifications of the rise of China and postunification security arrangements. In other words, while it may be difficult for outsiders to understand South Korea's complex weltanschauung and matching apprehensions, realpolitik is combined with national identity concern for Japan to come to terms with its wartime history. From Seoul's perspective, the importance of stressing a "correct understanding of history" remains a central imperative of its foreign policy narrative just as much as fostering a regional balance of power that remains favorable to its core security interests including the road toward reunification. Coping with a deeply ingrained security angst and emphasizing the importance of outstanding historical legacies are inseparable from this perspective.

In a May 2013 Asan Survey, when South Koreans were asked which countries posed a threat to South Korea, the results were as follows: North Korea (61.4 percent), China (59.7 percent), Japan (55.9 percent), and the United States (35.5 percent). In a July 2014 Asan Survey, the country that posed the greatest threat to South Korea was perceived to be North Korea (47.1 percent) followed by China (18 percent), Japan (14.5 percent), the United States (9.5 percent), and, most interestingly, Iran (0.3 percent) with 10.6 percent who did not know. Given such anxieties, it comes as no surprise that 89.6 percent of Koreans responded in a September 2013 Asan Survey that the ROK-US alliance continued to be needed and 83.9 percent replied that US forces were necessary for South Korea's defense. In a July 2013 Pew Research Global Attitudes Survey of perceptions of the United States and China, 78 percent of South Koreans had a positive view of the United States compared to only 46 percent for China. Interestingly, South Korea's favorability rating of the United States of 78 percent was demonstrably higher than Japan's 69 percent rating while only 5 percent of Japanese had favorable views of China. In addition, according to a September 2013 Pew Research Global Attitudes Survey on Russia, South Korea had the highest favorability rating (53 percent) of eight countries in the Asia-Pacific region (the others being Japan, the Philippines, Australia, China, Indonesia, Pakistan, and Malaysia).

As Park marked her first anniversary as president in February 2014, every major opinion poll showed that foreign policy was considered to be her strongest suit. An MBC poll conducted on February 25, 2014, revealed that Park's overall approval rating was 62 percent, with 67 percent who felt that her most notable achievement was in managing foreign affairs. Other polls also showed that Park's approval rating for foreign policy was between 60 and 70 percent. For example, a February 2014 Asan Survey revealed that 60.5 percent responded favorably to Park's policy toward the United States while 51.9 percent supported her policy toward Japan. In other words, while South Koreans feel very strongly about the vestiges of colonialism and remaining historical disputes with Japan, they are also aware that if bilateral ties worsen beyond repair, it will have major repercussions for South Korean security. For South Koreans, "correct historical perceptions" are as important as "security cooperation" in the context of Korean-Japanese ties.

In 2015, South Korea and Japan will mark the fiftieth anniversary of the signing of the 1965 Basic Treaty that normalized bilateral relation and both Park and Abe would benefit from forward movements. The Abe government has recently stated that it will continue to uphold the Kono and Murayama statements that are considered to be landmarks in the government's admission of the military's involvement in sexual slavery during the war years, but at the same time, it has asserted its right to secretly review the process of how the Kono statement was made in the first place. Such moves are considered by South Korea to be "doublespeak," the language that was used by Big Brother in George Orwell's *1984*. As a result, despite decades of the closest of economic linkages, wide-ranging cultural contacts, common strategic interests based on bilateral alliances with the United States, and shared democratic values, South Koreans today believe that a fundamental improvement in bilateral ties can only be undertaken when Japan accepts its wartime legacies.

In the final analysis, as much as alliances are driven and maintained by common interests and shared threat perceptions, they are also affected by varying degrees of values, historical experiences, and norms. The Korean-Japanese relationship is not an exception. For the vast majority of South Koreans, the legacies of colonialism such as the untold pain of the remaining sexual slaves who are all in their mid-eighties or older, the thousands of slave laborers, and the inhumane operations under Unit 731 are as pertinent today as they were in the war years. And if Abe and his supporters continue to whitewash history, South Korea cannot but continue to perceive such developments as Japan's version of apartheid, an internal policy so abhorrent that other countries felt justified in treating the transgressor as a pariah. Much has been achieved over the past half century in Korean-Japanese ties, and both continue to maintain the strongest of security ties with the United States. But alliance management in the context of ROK-US-Japan trilateral cooperation can no longer sideline historical legacies. It is time to relearn history and not to bury it under the carpet.

CHAPTER 12

Rebalancing and Entanglement: America's Dilemma in East Asia

Kiichi Fujiwara

Two years after Secretary Hillary Clinton's statement on America's refocusing on the Asia-Pacific region, vividly and controversially expressed as a "pivot" to Asia, we still do not see any significant increase of American influence in the East Asian region. China has maintained her tough position on territorial claims that has led to conflicts with neighboring Vietnam, the Philippines, and Japan, while the relations between South Korea and Japan—the principal American allies in the region—have stalled since the inauguration of President Park Geun-hye.

How could this be? The explanation lies in the inherent dilemma entailed within that rebalancing policy: Balancing alone cannot change the behavior of nations where strong policy commitments rooted in domestic politics make concessions to the adversary all but impossible. In the following, I will offer a brief sketch of the rebalancing policy and its limits, in order to understand the poor consequence that we observe so far.

Policy of Rebalancing

In her article, Clinton highlighted the Asia-Pacific region as vital to the economic and strategic interests of the United States, and urged an American "pivot" to the region.[1] Her focus on the need to keep open markets and freedom of navigation, although not new as such, implied a stronger awareness of the potential risk that may accompany the rise of China, accompanying a call for strengthening US alliances and partnerships. Secretary of Defense Leon Panetta echoed Clinton's assertions in his remarks at the Shangri-La Dialogue in June 2012, where he advocated new investments in capabilities that are needed to project power and operate in the Asia Pacific to advance peace and security in the region.[2]

The background to rebalancing was only too obvious. During the more than one decade of American involvement in the "war on terror," the rise of China and the expansion of naval activities in blue waters had intensified territorial disputes between China and Vietnam, the Philippines, and Japan. Although Washington has been traditionally reluctant to be involved in the controversies over territorial control, the use of force to advance Chinese interests was clearly against the maintenance of the status quo in the region, thus crossing the terrain of border conflicts and presenting a geopolitical crisis at the regional level. With troops returning from Afghanistan and Iraq, the Asia Pacific was a natural priority for the United States to rebuild her deterrence capabilities and exert influence through the network of alliances and partnerships so that China will refrain from adventures that may destabilize the region. It was good policy for maintaining regional stability, American interests, and credibility to US allies.

The policy was balancing and not war-fighting. After a prolonged period of sending troops to two fronts, Washington was not keen on fighting another war; combat with Chinese troops was far from what many were prepared to contemplate. Thus, the focus on alliance and partnerships; the pivot would depend on strengthening existing alliances with Australia, South Korea, and Japan, and also seeking the possibility of "mini-laterals," the subset of partnerships involving two or more US allies so that the hub-and-spokes network of bilateral alliances would develop into a more multilateral network along the lines of NATO in Europe. It was expected that partnerships, such as the Japan-Australia Joint Declaration on Security Cooperation of 2007, would provide models for strengthening defense cooperation between South Korea and Japan.[3]

Limits of Rebalancing

Initial concerns over this pivot focused either on the nature of Chinese reactions, which themselves may disrupt regional stability, or on the prudence of shifting attention to Asia at a time when the instability in the Middle East still remained alarming.[4] There was, however, yet another limit to the rebalancing: this strategy will not work when you wish to avoid entanglement, a dilemma that surfaced with Japan's nationalization of the Senkaku (Diaoyu) islands in September 2012, which was most certainly not a challenge to the status quo, as the islands had been under Japanese control since the reversion of Okinawa in 1972. Tokyo governor Ishihara Shintaro, known for his extreme views toward China, had declared his intention to buy the islands from private owners, supported by a donation campaign that collected 1.4 billion yen. Nationalization was a reaction to stop Ishihara's move. The irony is that the scheme intended to avoid a provocation ended up provoking the strongest anti-Japanese demonstrations and riots, some of which were accompanied by ugly violence. Neither Beijing nor Tokyo was preparing for war, but neither was ready to back off, leading to a real possibility of a military incident over the ownership of a couple of rocks.

Facing this development, Washington turned to a more cautious position that asks moderation of both China and Japan. Newly appointed secretary of state John Kerry visited both Beijing and Tokyo in April 2013 but failed to reduce tensions. It was Foreign Minister Kishida Fumio, not Kerry, who used the word "rebalancing" in the press conference.[5]

With the Chinese and the Japanese strongly committed to their respective causes, it was clear that the Senkaku island crisis could not be resolved by US diplomatic influence. Rebalancing, after all, is yet another variation of the classical balance of power policy that aims to challenge the behavior of a potential adversary and change it to behavior that matches the interests of the balancer. If the adversary operates according to rational calculations of costs and benefits, balancing may well work out. If the adversary, however, is committed to a certain policy to the extent that it is willing to risk the possibility of major aggression, balancing will fail, leading to a choice between doing nothing and going to war.

In the case of the Senkaku islands, I doubt if either side is ready to go to all-out war. It is precisely the ambiguity of escalation, however, that allows this game of chicken to continue. If Washington is not eager to risk a potential war with China, a tiny conflict such as the one around the Senkaku islands may prove to be sufficient to make Washington reluctant to pursue a policy of rebalancing and restrain itself from further commitment.

Collective Defense

The shift from rebalancing to moderation was evident in regard to collective defense. Japan had continuously claimed that collective defense is unconstitutional, ruling out military engagement that is necessary to carry out alliance obligations. As this severely limits the use of Japan's forces as an American ally, the US government has continuously pressed Tokyo to change this position and find ways by which Japan could work together with US forces as a "normal" ally. One would imagine that a pivot to Asia would increase the need for more active defense cooperation between Tokyo and Washington, and that a government in Tokyo changing its constitutional interpretation and allowing collective defense would be welcomed. But Washington, while supporting the new Japanese National Security Council and welcoming Japan's acceptance of surveillance drones in Japan, was rather cautious in pressing reforms toward the acceptance of collective defense, even though the Abe administration has listed that as a priority.[6]

Of course it is not that Washington does not want Japan to accept collective security. A new understanding of the Constitution will widen the scope of US-Japanese defense cooperation, adding power to US initiatives without additional financial costs. There was, however, the question of timing: When will be the appropriate time for Japan to move forward? A Japanese government declaring that it accepts collective defense was sure to be seen as opening the Pandora's box in Asia, lifting the lid that may have restrained Japan's military action. Such reform could provoke strong reactions from both China and South Korea, perceived as a return of Japan's "militarism," and further accelerate tensions in the region; not what the United States wanted.

Alliance politics necessarily arouse two anxieties, abandonment and entanglement. Fear of entanglement, where Japan will be forced to join a war started by the United States, was the central cause of opposition to collective defense in Japanese politics. Entanglement, however, can work both ways. A military conflict that involves a junior partner of an alliance places a nation in the unenviable position of becoming involved in a conflict that may produce consequences against its interests. It was only natural that Washington would be cautious about becoming the dog wagged by its tail.

History and Wartime Responsibility

Another issue that has reduced the impact of Washington's rebalancing policy was the debate over Japan's responsibility over war and colonialism, which became particularly intense with the arrival of the Abe administration. One has only to see the refusal of the Korean president to meet Abe face-to-face to understand how seriously the past can endanger the present. Unlike freedom of navigation and maritime safety, war memory (and history) is an issue that is sure to isolate Japan in East Asian politics, but that did not prevent Abe from visiting the Yasukuni Shrine in December 2013. Abe has been an active member of the conservative group that openly denounced the Tokyo International Tribunal as victor's justice, the Nanjing massacre as a myth propagated by the Chinese Communist Party, and sexual slavery as mere prostitution twisted into a crime by the Japanese left and Koreans. For the conservatives, this is a movement that would restore the proud history of Japan against what they see as distortions propagated by China, South Korea, and the left in Japan.

From the Chinese or the Korean perspective, such conservative reading of history is precisely what makes it impossible to work with the Japanese. If there are committed ideologues in Japan, there are also committed groups in China and South Korea that refuse to drop their position in the name of regional cooperation. War memories here have enhanced nationalization of politics in East Asia, where nationalist reading of history reduces the possibility of pragmatic conflict resolution.

The history debate further eroded the influence of American rebalancing, especially since the debate was particularly intense between South Korea and Japan, both allies of the United States. Washington could, of course, make a strong statement denouncing what many Americans see as the whitewashing of wartime atrocities, but then such a reaction might further push Japan to the right; if Washington ignored historical revisionism in favor of pragmatic conflict resolution and tried to force South Korea to work with Japan, the alliance with South Korea would be endangered by angry Korean citizens. In the face of strongly committed public opinion in favor of confrontational policies, there is little that can be done. Washington now not only must fear entanglement in regional territorial disputes, but also historical disputes where taking either side is sure to work against her interest.

Conclusion

Considering the size of the regional economy and the changing balance of power in East Asia, the pivot to Asia was a necessary and clever move for the United States to reformulate her foreign policy after the wars in Afghanistan and Iraq. Balancing, however, cannot be expected to be productive when the major power is reluctant to be further involved, or entangled, in regional disputes; it is also almost impossible to succeed when participating powers refuse to change their policies. One successful case of balancing took place in late-nineteenth-century Europe, where Count Bismarck worked on a careful diplomacy to draw both Russia and Austria to his side. That balancing did not work out, however, due to irreconcilable conflicts of interest between Russia and Austria, paving the way to the political polarization of Europe in the early twentieth century.

Abe and Park have finally agreed to attend a summit meeting, along with Obama, in The Hague this March. This summit was made possible through Washington's continuous and patient diplomatic initiative to the two nations. But it is still premature to hail this achievement, for mutual suspicion between Japan and South Korea has not receded at all. We have yet to see whether the Obama administration will escape the failure of rebalancing, as illustrated in the erosion of balance of power politics in Europe.

Notes

1. Hillary Clinton, "America's Pacific Century," *Foreign Policy*, November 2011.
2. Full text in: http://www.defense.gov/speeches/speech.aspx-speechid=1681.
3. http://www.mofa.go.jp/region/asia-paci/australia/joint0703.html. For a more detailed discussion, see William Tow and Rikki Kersten, eds., *Bilateral Perspectives on Regional Security: Australia, Japan and the Asia-Pacific Region* (New York: Palgrave Macmillan, 2012).
4. See, for example, Ely Ratner, "Rebalancing to Asia with an Insecure China," *Washington Quarterly* 36, no. 2 (2013): 1–38. Aaron Friedberg discusses rebalancing as a combination of engagement and balancing in "Bucking Beijing: An Alternative U.S. China Policy," *Foreign Affairs* 91, no. 5 (2012): 48–58.
5. http://www.state.gov/secretary/remarks /2013/04/207483.htm.
6. *The New York Times*, October 13, 2013.

CHAPTER 13

President Obama and Japan-South Korean Relations

Sheila A. Smith

President Obama's trip to Asia in April 2014 will have a full agenda, and should result in some important accomplishments. He will have his work cut out for him, however. The region's challenges are far deeper and the solutions to problems far more complex than most Americans realize. One trip alone cannot address all the issues that demand the president's attention, but this trip will strengthen some very important relationships—and stimulate some constructive attention to a diplomatic logjam between Tokyo and Seoul.

The April trip includes four destinations, each with its own agenda for the US president. Two of his stops are to countries that were expecting him last fall, when he was due to participate in the East Asia Summit. A US president had not visited Malaysia for nearly 50 years; so it was particularly wrenching when politics in Washington took priority. In Kuala Lumpur, Prime Minister Najib Razak seeks to transform his nation and instill in its economy a new embrace of competitiveness and global reach. He faces domestic pressures on his effort to conclude the TPP, and the president's visit will help focus attention on the benefits to be had from his transformative agenda. It will also give the two leaders a chance to discuss the growing pressures Malaysia is facing from Beijing.

In Manila, President Benigno Aquino will be particularly keen to have Obama's support as his government continues to confront Chinese pressure over their maritime and territorial disputes. Manila's decision to revamp its Visiting Forces Agreement (VFA) with the United States comes from its own difficulties with Beijing, and undoubtedly Aquino will be looking to Obama for support in broadening his diplomatic effort to manage China's economic pressures on the Philippine economy. Expectations are high for what a visit from Obama can bring to both of these Southeast Asian economies.

The US Role in Reconciliation Diplomacy

The president will start his visit in Northeast Asia with stays in Tokyo and Seoul, and it is here that the more difficult challenge resides. Washington's bilateral relations with its two allies are quite strong, but it is their relationship with each other that has suffered over the past 18 months. Popular sentiments have become increasingly sensitive as diplomacy between President Park Geun-hye and Prime Minister Abe Shinzo stalled. Obama and his staff have worked hard with colleagues in Tokyo and Seoul to lay the groundwork for a leaders' summit at the Nuclear Security Summit next week at The Hague. Given the theme of that meeting, it is likely that the three leaders will focus their discussions on North Korea and on steps to coordinate their policies for coping with nuclear and missile proliferation. Washington has an interest in discussing regional cooperation in ballistic missile defense and other aspects of security cooperation.

Obama's meeting with Abe and Park demonstrates his commitment to facilitating communication between the two US allies, and the time between the Nuclear Security Summit and the president's visit to the region in April should offer an opportunity to work through some of the issues that caused their estrangement. The momentum of the trilateral next week ought to serve as encouragement for both Park and Abe to tackle head on some of their outstanding differences. But it will be important to recognize the causes, the consequences, and the role of the US president in facilitating better relations between Tokyo and Seoul.

The Causes of Japan-Korea Estrangement

The perceived cause of the estrangement is different in each capital. In Seoul there is a growing sense that the Japanese government is reneging on past statements of remorse for World War II, and Abe's ambiguous statements about the past in the early months of his return to power raised questions about his commitment to the Japanese government's official positions on the past: the Murayama statement of 1995 and the

Kono statement of 1993. Moreover, by last fall, there was a growing worry in Seoul that Japan's defense reforms would somehow impinge on Korea's own security, a worry that seemed particularly misplaced when it was focused on the US-Japan Defense Cooperation Guidelines review announced last October at the Security Consultative Committee (2 + 2) meeting in Tokyo between Secretary of Defense Chuck Hagel and Secretary of State John Kerry and their Japanese counterparts, Minister of Defense Onodera Itsunori and Ministry of Foreign Affairs Kishida Fumio.

The tensions, of course, predate Abe and Park. In Tokyo, President Lee Myung-bak's ill-fated visit to Dokdo Island (Takeshima) in the summer of 2012 compounded a growing unease about the trajectory of Korea's domestic politics. Activism related to Japan's colonization of the Korean Peninsula focused attention on an array of issues, including a court case over the sexual slavery of Korean women during the war as well as legal claims for compensation for forced labor under Japanese companies. The issues were not new, but the mechanism of protest puts new pressures on the South Korean government. The use of the courts to adjudicate victim compensation and to compel the government to advocate for those who felt Japan needed to do more to demonstrate its remorse, despite the bilateral treaty of 1965 and the discussion of compensation that accompanied it, came due to new generations of South Koreans pressing their government for greater accountability and a more forceful representation of their interests in negotiations.

Leadership transitions in both countries also contributed to the growing frictions in 2012. As the presidential election approached, Lee found himself increasingly confronting a contentious legislature. Likewise in Tokyo, Prime Minister Noda Yoshihiko found his government constrained by a "twisted" parliament, with opposition critics in the Upper House particularly confrontational. Elections in December 2012 produced a chance to reset the relationship, but unfortunately the initial diplomatic overtures just raised more hackles. In February 2013, Deputy Prime Minister Aso Taro's visit with newly elected president Park did not go well, and later efforts to establish contact with the Blue House were unable to undo the damage.

Park was clearly focused on her relationship with the leaders of the United States and China. Early in her tenure, she traveled to both capitals and was received by appreciative audiences. Abe, in contrast, received a cold shoulder from the new Korean president and was also confronted by the legacy of the downturn in Japan-China relations that resulted after his predecessor purchased the Senkaku islands, inflaming

Chinese popular outrage and anti-Japan demonstrations across the country. Park's obvious interest in her relationship with Xi Jinping, and her lack of interest in a dialogue with Abe only served to exacerbate the relationship.

Protracted diplomatic silence over this past year has allowed the popular feelings in both countries to decline precipitously. Public opinion polls have revealed a striking drop off in trust between the two countries. Indeed, in South Korea, the Asan Institute revealed that Kim Jong-un seemed to rank higher in the Korean public's estimation than Japan's prime minister. In Japan, outbursts of anti-Korean demonstrations shocked the nation as hateful speech and discriminatory language poured from the vocal minority offended by South Korean actions. A counter effort by citizens who opposed this kind of discriminatory behavior toward Koreans also emerged, with volunteers confronting those with an anti-Korean bent and erasing hateful graffiti from public spaces. With no dialogue between the two leaders, the political opportunity for antagonism and nationalism was allowed to grow in both countries.

Ties between Japanese and South Korean nongovernment groups remain strong, but the political frictions are beginning to take their toll. Although Japanese companies continue to have good relations with their Korean partners, a drop in economic activity between Japan and South Korea was visible in 2013. Trade between them took a slight dip in 2012, but rebounded in 2013. Total trade grew 5.96 percent from 2010 to 2011; shrank 4.35 percent in 2012; and grew 10.55 percent in 2013.[1] FDI was strong through 2012, but Japanese investment in South Korea shrank in 2013. It grew from USD 1.085 billion in 2010 to USD 2.439 billion in 2011 to USD 3.996 billion in 2012, before shrinking approximately 17.8 percent in 2013 to USD 3.286 billion. South Korean investment in the Japanese economy shrank an alarming 91.6 percent in 2013, down to USD 47 million from a high of USD 559 million in 2012.[2] The prospect of a court judgment in favor of redress for forced POW labor by Japanese companies, however, could have a serious impact on business relations between Japan and the ROK.

The President and US-Japan-ROK Summitry

From last year, the costs of this estrangement between Seoul and Tokyo became a focal point in Washington's diplomacy toward both countries. In large part, however, the consensus was that the US government could not broker reconciliation between them. Only Park and Abe could do that, and yet, the worsening relations and the frustrations that were brewing between the governments became troubling to US policymakers. The costs seemed all too obvious: a belligerent and adventurous North Korea could easily upset the region again.

More difficult was the growing criticism from Beijing and Seoul of Tokyo. Diplomatic isolation of Japan served no one's interests and shook the premise of stability that informal strategic cooperation between Japan and South Korea offered the region. By the end of 2013, the actions of both Park and Abe suggested worsening rather than improving relations. For over three months now, the US challenge has been to facilitate a different approach, and to find a way, at the highest level, to lead a trilateral dialogue on shared interests. That hard work has paid off, and next week the three leaders will meet on the sidelines of the Nuclear Security Summit. Obama's participation in the meeting offers the opportunity for Abe and Park to meet face-to-face and to demonstrate their political commitment to repairing their damaged ties. No less important is the demonstration at that summit of Obama's political commitment to ensuring continued effort at improvement.

For many, this may seem to be a photo op—designed to paper over differences—but the significance goes far beyond that. Last week, Abe in the Japanese Diet clearly stated that he would support the Murayama statement and, most importantly for Park, the Kono statement. Leaving no room for ambiguity about his intentions, he confronted those in the Japanese parliament who had already begun to demand an investigation into the Kono statement—how the evidence it relied upon was produced and even the reliability of the statements from the victims of the military brothels. This investigation will proceed, but the Abe Cabinet has committed itself to maintaining the statement of remorse issued by Cabinet Secretary Kono to the women forced into sexual slavery during the war.

For its part, the Korean government now will need to consider its own way forward. Park has advocated a "correct understanding of history," yet the practical steps for a reconciliation strategy remain vague. She too has publicly stated that the time has come to improve South Korea's relations with Japan, and she may in fact have the support of the Korean people behind her in that objective. Her participation in a trilateral summit indicates her commitment to act.

April and Beyond: The Parameters of Reassurance

Obama has played a crucial role in helping Park and Abe find their way to a summit meeting, and this is no small achievement. From here on out, the question continues to be how to reestablish a firm foundation for the trilateral conversation among Japan, South Korea, and the United States. North Korea, nonproliferation, and regional security (including trilateral military cooperation) offer an agenda that reflects their shared security goals. This will be more than enough for The Hague meeting.

What comes afterward remains to be seen, but the process has begun with the commitment to meet. Already the minimum requirements of Park have been met. Abe has reaffirmed, clearly and without reservation, the Murayama and Kono statements. His government has also reassured both Washington and Seoul that the Kono statement will not be revised, no matter the sort of procedural review conducted in response in the Diet. The other side of the equation now is for Park to reassure Tokyo that her government has no intention of reopening discussions on wartime compensation, a discussion that both governments completed in their 1965 peace treaty. These two bookends should define the parameters of reassurance needed for Tokyo and Seoul to proceed to consider what other practical steps might be taken to demonstrate their mutual commitment to deepening their ties.

As he looks ahead to this conversation, Obama should not hesitate to consider the history of reconciliation that successive US governments have forged with Seoul and Tokyo. US willingness to own up to past mistakes has been tested in both relationships. One of the pillars of strength in the US relationship with Japan continues to be historical reconciliation. While the United States cannot define the terms of reconciliation between others, Obama can encourage and support the objective itself. He can be the facilitator-in-chief, and should not underestimate his ability to appeal not only to Abe and Park but also to the citizens of both of their countries. His April visit offers just such an opportunity. It may take time, but the United States will continue to stand by both of its allies in Northeast Asia as they attempt to find a lasting and sustainable vision for reassurance and reconciliation.

Notes

1. Data are from MOF Trade Statistics, http://www.customs.go.jp/toukei/info/index_e.htm.
2. Korean FDI to Japan was USD 274 million in 2010, USD 197 million in 2011, USD 559 million in 2012, and apparently shrank 91.6 percent in 2013 to only USD 47 million. FDI from JETRO (in dollars), https://www.jetro.go.jp/en/reports/statistics/; 2013 numbers are provisional.

CHAPTER 14

State of the Rebalance

Thomas Hubbard

President Obama's recent trip to Asia was a timely, perhaps necessary, reminder that the "rebalance" to Asia remains an important pillar of US foreign policy despite the distraction of immediate crises elsewhere in the world. During an unusually lengthy eight-day trip, Obama visited three treaty allies—Japan, Korea, and the Philippines—and a long-neglected friend, Malaysia. Reacting to the first stop in Japan, US media tended to focus on what the trip did not achieve—a clear-cut breakthrough toward the Trans-Pacific Partnership (TPP), the central economic component of the rebalance. That was unfortunate, but the real message of this trip was security, and the principal audience was allies and friends seeking assurance that the United States will stand by its commitments at a time of profound regional change. The visit was also intended as a strong signal to China, whose assertive actions across a broad front have shaken the region. The strong security thrust resonated well in all of the countries visited and produced concrete results that will strengthen the US ability to contribute to regional security. It remains to be seen whether this strong demonstration of allied solidarity will have a positive effect on China's behavior.

Japan: Enhanced Alliance, but Too Soon for Trade Deal

An Obama visit to Japan was long overdue. As one who participated in President Clinton's 1996 visit to Japan, I was surprised to realize that this was the first state visit by an American president since then (an anomaly caused largely by the frequent turnover in Japanese leaders).

This trip was a timely opportunity to build ties with Prime Minister Abe Shinzo, who is likely to be in office for several more years, and to reinforce his efforts to upgrade Japan's defense posture. The shadow of an assertive China loomed large over the talks, as it did elsewhere on the trip. The keynote was Obama's clear indication that the United States considers the disputed Senkaku islands, long administered by Japan, to fall squarely under the defense commitments embodied in our Mutual Security Treaty. Successive secretaries of state and defense have voiced this commitment, but this is the first time this commitment to come to Japan's aid in the event of an attack against the Senkakus was explicitly stated by a president. With an eye toward China, Obama was careful to say the United States takes no stand on ultimate sovereignty over the islands and to urge both sides to exercise restraint, but the message was clear. Obama also took the occasion to support steps Abe is taking to enhance Japan's ability to contribute to the military alliance. Recognizing that both China and South Korea have chosen to portray these steps as a return to militarism, Obama urged Abe to move forward in a transparent manner that does not cause tensions with neighboring countries and made clear the US wish that Abe avoid symbolic steps, such as further visits to the Yasukuni Shrine, that would exacerbate tensions.

The US side wanted more concrete economic results. In the run-up to the visit, US officials did little to hide their hope that a breakthrough on market access for agricultural products and automobiles would give a boost to the broader TPP negotiations. Recognizing that the US-Japan talks are in many ways the main event, other TPP parties have been awaiting two critical developments: (1) US steps toward securing Trade Promotion Authority (TPA) from Congress; and (2) achievement of a US-Japan deal on marker access for automobiles and agricultural products. The United States was hopeful that a breakthrough with Japan would make it possible to overcome widespread opposition to TPA on the part of congressional Democrats; it gambled that TPP was so central to Abe's "three arrow" program of economic reform that he would make the necessary tough decisions on agricultural imports even in the absence of assurance that Congress would ultimately back TPP. That gamble failed, neither leader being prepared at this stage to compromise domestic interests. Following talks that took up the better part of Obama's time in Japan, US officials tried to project optimism, asserting that the two parties were able to lay a pathway for eventual success and thereby inject fresh momentum into the larger TPP talks. Talks continue both bilaterally and multilaterally, but the failure to achieve a deal in Tokyo did little to dispel the widespread view in Asia that there will be no TPP deal until after the US midterm elections.

Korea: A Necessary Gesture

The White House did not initially plan a stop in Korea, since Obama had already visited Korea three times and also hosted President Park for an official visit to Washington, but Obama wisely decided to add Seoul to his itinerary, recognizing that a stopover there would underscore solidarity in dealing with North Korea and, perhaps, add momentum to his efforts to reduce tensions between Japan and South Korea. As it transpired, the visit offered an opportunity to express solidarity with a Korean public profoundly racked with remorse over the death of more than 250 high school students in a ferry accident. Koreans were deeply moved by Obama's expressions of sympathy.

North Korea did its part to add excitement to the visit. Obama arrived in Seoul just after a series of North Korean short- and medium-range missile tests and amidst new signs of preparations for a fourth underground nuclear weapons test. Against this backdrop, the two presidents focused largely on affirming the US-ROK alliance. Among other things, Obama agreed to a long-standing Korean request to review a decision by their predecessors to turn wartime command of combined forces to a Korean general. In order to reinforce the importance of the joint posture, Obama and Park also conducted the first joint presidential visit to the Combined Command at Yongsan Garrison since its establishment 40 years ago. The North Koreans reacted with vituperative blasts at both leaders for preparing for war.

Malaysia: Finally a Presidential Visit

No US president had visited Malaysia, an important trading partner, since Lyndon Johnson; both Clinton and Obama had postponed scheduled visits. As with his culturally appropriate condolence gestures in Korea, Obama displayed his affinity for Malay culture (he actually speaks the national language the Malays share with Indonesia) to charm a proud public still stung by negative international attention to its government's poor handling of the Malaysian Airlines disappearance.

Although the two states have long engaged in quiet defense cooperation, there was a time when Malaysia would have been uncomfortable with being grouped with three US treaty allies. Yet, both sides are now describing their ties as a "comprehensive partnership," a new phase in relations that is possible in part because Malaysia wants deeper and more rewarding economic ties with the United States, but also because Kuala Lumpur sees the value of America's security role in the Pacific. Malaysia's decision to join the Proliferation Security Initiative as well as

Prime Minister Najib Razak's strong support for TPP despite considerable political opposition at home were important concrete outcomes of this long-overdue visit to Malaysia.

The Philippines: Toward a More Functional Alliance

Obama had also planned to visit the Philippines at the time of the government shutdown late last year, a visit that was originally to highlight its recent success in combating corruption and promoting economic growth—at around 7 percent, the highest in Southeast Asia over the past couple of years. Since then, the Philippines has grappled with the effects of a devastating typhoon and become increasingly concerned over Chinese incursions into disputed islands and waters. Thus, the security focus of the visit was welcomed by both Manila and Washington. In the most concrete result of the entire Asia trip, Obama and President Aquino presided over the conclusion of a new military agreement, the Enhanced Defense Cooperation Agreement (ECDA), which, among other things, will allow regular rotation of US military personal through US-controlled facilities within Philippine military bases. Long under negotiation, ECDA will facilitate the most significant presence of US forces since the Philippines kicked the United States out of the Clark and Subic bases in 1992 and send a clear signal of the US intent to help it develop greater ability to defend its interests in the South China Sea.

Unlike the case with Japan and the Senkakus, the applicability of the US-Philippines Mutual Defense Treaty to the Philippines' claims in the South China Sea is less than finite, and Obama pointed out that the United States takes no stand on sovereignty over the disputed islands. However, he sent a clear signal to the Chinese that it is not neutral when it comes to provocative steps affecting allies in disputed territories. To underscore that point, Obama endorsed the Philippine decision, strongly opposed by China, to seek a ruling on China's entire "nine-dash line" claim in the South China Sea by the UN Law of the Sea tribunal.

Impact on China

The presidential visits as a whole revealed, without a doubt, that China's assertion of ambitious territorial claims in the region has rattled smaller neighbors and consolidated determination in countries to make the most of the US counterbalance to China. While Obama tried to show balance at each stop, the security-dominated visit may have strengthened the widespread misperception among officials and the public in China that the United States is manipulating Asian countries, Japan included, to contain China's modernization and deny China what it considers to be its rightful place in the world. Although China appears prepared to adjust tactically to the ebbs and flows of its maritime disputes, particularly with regard to the most dangerous dispute with Japan, its strategic direction is clear: relentless pursuit of expansive territorial claims and rejection of international standards, including binding arbitration as a means of peacefully resolving such disputes. A recent regional agreement on rules of the road for naval encounters is positive, but there is scant evidence that the Chinese are willing to accept the nuanced position Obama delivered during the visit. Rather, Beijing is likely to continue to look for opportunities to consolidate its claims, as it did by towing a heavily guarded oil rig into waters disputed with Vietnam. That provocative move led to violent encounters with Vietnamese vessels and deadly anti-Chinese riots in Vietnam, which destroyed a considerable amount of Korean, Japanese, and Taiwanese property. The ramifications of this outburst will keep maritime claims in the South China Sea on the front burner for some time to come.

The Future

Stasis in the TPP talks, combined with preoccupation with the China challenge, made it inevitable that this trip would be dominated by security developments. Yet, in the longer term, success of the rebalance to Asia will depend not just on security pledges but equally importantly on confidence in our capacity to carry them out. It was fortunate that this trip finally took place at a time when the austerity mood in Washington is palpably diminishing, particularly with regard to defense spending. Asian confidence in US defense commitments is also heavily dependent on how it fulfills commitments and exercises leadership elsewhere in the world. In this sense, it was a healthy signal to Asians that the president both made his promised trip to Asia despite the pressing crisis in Europe and was seen in each capital he visited to be actively engaged in leading the international response to Russian aggression. Finally, the rebalance to Asia will not succeed without a strong economic component, and TPP is the central element. The TPP negotiations may not go far before the midterm elections, but it is important that Obama thereafter engage his personal prestige and energy to achieving this agreement that will lay the basis for US economic engagement with the world's most dynamic region.

CHAPTER 15

US-Japan Relations: Do We Share the Same Values and View of History? US, Japanese, and South Korean Perspectives

Dennis Blair, Nakanishi Hiroshi,
Bong Youngshik, and Hyun Daesong

Part 1: A US Perspective

Admiral Dennis Blair

History plays an important, changing, and, often, elusive role in the lives of nations. It is important because of its strong influence on the political life of a country. The fourth century BC Greek historian Thucydides presciently explained that nations go to war because of honor, fear, and interest. A country's history is inextricably bound up with its honor, and often with its fears. History's role in a nation's life can change because history is not a fixed set of facts understood in the same manner by all forever. Fresh experiences recede into memory; new facts about old events come to light. Even more important, the same facts may be reinterpreted by new observers, changing significantly the nature and magnitude of their impact on public opinion and politics. The role of history is also elusive because there are times in a nation's life when the impact of historical events looms large, and other times when it does not.

History affects both Japan and the United States in important, changing, and subtle ways, but there are differences between the two countries. Some of the differences are easily explained. Japan is a homogeneous nation. Its people have lived on their island country for centuries with little immigration and, except for the past century and a half, little interaction with the outside world. Japanese look to a single history, beginning before recorded time, based on myths and legends, many of them carried through to recent times. The United States is a land of immigrants who bring their own views of history with them to their new home, but, upon arrival, have experiences that often temper those views. All of American history, save that of the Native Americans, took place in the past 400 years with written records to document it. It is little wonder that Americans are used to contending views of history, to evolving and new interpretations of old events, whereas Japanese have been used to a more singular narrative of their past.

Another difference between the two countries has been the role of history in governance. Successive Japanese governments until the end of World War II used history for dynastic purposes. They held huge public ceremonies to celebrate the founding of the empire in 660 BC—despite scant evidence of the accuracy of these formative events to support this story. The United States has its legends too, and American leaders have invoked mythic views of history for governing purposes. But there are written records to support or debunk those myths, and new books and movies constantly provide reinterpretations of, for example, the Founding Fathers and the Wild West.

The way history is taught, learned, and understood is also different in the two countries. Japan has a unified education system across the country. There is one national history curriculum for all Japanese students. Moreover, that curriculum has a single interpretation of Japanese history that students are required to remember and recite in tests. In the United States, education is a local responsibility, and textbooks and teaching emphases are different across localities. In an advanced placement American high school history class examination, primary sources will be provided with the examination, and students are expected to use those sources to provide their own interpretations of historical events.

Historical reinterpretation is common in the United States, both in academia and in popular culture. New books often provide entirely different views of periods in American history; new movies put a different cast on well-known events or popularize events formerly little known. Japan has a much less freewheeling discussion of its history. In the past, it was confined by censorship, but even today discussion of difficult topics may be restricted by cultural norms. One of the most important recent books about Japanese history, *Japan 1941*, written by a Japanese scholar working in the United States, is little known in Japan. The most popular Japanese historical movies, the samurai epics, do not break new historical ground. American movies such as *Lincoln* or books like *The Immortal Life of Henrietta Lacks* bring entirely new aspects of history to the public's attention, some of it positive and some negative.

So domestic history is used, taught, and developed differently in Japan and the United States. What about our common history? In the living memory of older Japanese and Americans, the two countries went to war with one another, and the history of how that war began, how it was fought, and the events that followed it form our common history; but this common history is often interpreted quite differently in the two countries.

Views on the start and the end of the war, in particular, diverge widely. Americans generally consider the Japanese surprise attack on Pearl Harbor an unjustified act of treachery, made despite ongoing good-faith negotiations. They know little about the background of anti-Japanese legislation in California in the early part of the twentieth century or American actions to cut off raw materials to Japan in the 1930s. The official Japanese interpretation of Pearl Harbor, meanwhile, has been sanitized. A brochure about Japan handed out to visitors to Japan's consulate in Honolulu in the 1980s described the beginning of the war as follows: "When Japan entered the war against the coalition led by the United States."

Most Americans believe that the decision to drop atomic bombs on Japan in August 1945 was fully justified by the cold calculation that the alternative, an invasion of Japan, would result in massive casualties for American and allied troops, as well as larger losses of Japanese lives, than would result from the bombings. Most Japanese focus more on the fact that the United States inflicted on Japan alone the most terrible weapon invented by mankind.

In addition to these very different understandings of the start and end of the war, there were other actions by both sides that caused deep resentment: America's unrestricted submarine attacks on all shipping to and from Japan and the firebombing of Tokyo; Japanese treatment of American and other prisoners of war.

Probably most important and regrettable, official propaganda in both countries portrayed the other in the most prejudiced, simplistic, and negative terms. American slang for Japanese soldiers was demeaning and racist; Japanese soldiers were taught that their American enemies did not have the fighting spirit and honor of Japanese, and that those who surrendered should be treated as slaves.

Given this background, it is extraordinary how positive relations between the two countries have been, starting from the day the war ended. The American occupation, which included the decision not to prosecute the emperor, the provision of economic assistance, and the political reconstruction of the country, generally produced gratitude rather than resentment among most Japanese. Japan's immediate cessation of hostilities at the command of the emperor, the cooperation with occupation authorities, and the stoic fortitude showed by citizens in rebuilding their country all engendered American admiration and respect. Within months of the end of the war, the Soviet Union's alignment against American and Japanese national interests both stimulated cooperation and restrained historical animosity, attitudes that continued throughout the decades of the Cold War and beyond.

Once the United States had prosecuted Japanese war criminals early in the occupation, there were few public disputes over historical issues. However, there were also few official or unofficial efforts to resolve such disputes or to better understand history from the point of view of the other side. Subsequent work by historians that has shed new light on the war, its prologue, and aftermath has come mostly from the American side, with new books that include research based on Japanese sources. Revisionist books written by Japanese have been less frequent, but often more controversial, especially those that have been critical of Japanese actions.

In popular culture, one unique but very positive phenomenon was the pair of 2006 movies directed by Clint Eastwood about the battle of Iwo Jima. The first, *Flags of Our Fathers*, was presented from the American point of view, while the second, *Letters from Iwo Jima*, told the story from the Japanese point of view. The story from the Japanese point of view received more popular attention in the United States than its companion movie; disappointingly, the reverse was not true among Japanese audiences. However, Japanese movies have made contributions to an evolution in popular understanding of the war. Movies about kamikaze pilots had for years idolized the sacrifice of the young aviators. An example was the very popular *For Those We Love*, released in 2007. Although this year's blockbuster on kamikaze pilots, *The Eternal Zero*, admires their patriotic determination to die for their country, it also depicts more of the young pilots' misgivings about their mission.

The treatment of history in textbooks and in popular culture and government attempts to manipulate and control history, however, are less influential than they would seem in forming popular attitudes. There is no doubt that in small minorities in both Japan and in the United States bitter resentments and extreme nationalistic views lie beneath the surface. However, what is more remarkable is the degree of respect and admiration between Americans and Japanese, despite the savageness of the war they fought.

According to this year's Global Attitudes Project by Pew Research, for example, 70 percent of Americans had favorable views of Japan, the highest rating among major Asian countries and double the percentage of those with a favorable view of China. Within Japan, 66 percent had a favorable view of the United States, again, the highest favorable rating of any other major country. These ratings demonstrate that the two countries, despite quite different national approaches to history, a brutal war, and minimal common attempts to resolve historical issues, have still developed mutual respect, understanding, and even affection. These favorable views have been achieved through the many individual points of contact between Americans and Japanese, the shared alliance of the past 70 years, and knowledge gained outside of the study of history.

However, a better understanding of history in both countries would lead to an even stronger relationship and would provide a sturdier foundation when setbacks and difficulties occur. It is therefore important that both countries continue to develop a fuller and more balanced understanding both of their individual and common histories. It is the responsibility of scholars, movie directors, think tank experts, bloggers, prominent citizens, and governments in both countries to resist simplistic interpretations of history, provide a straightforward understanding of the actions of both the heroes and the villains, and portray events in the context of the time they took place. Only with a solid and balanced understanding of their past can countries move into the future in a way that gives confidence to their own citizens and to those of other countries.

Part 2: A Japanese Perspective

Nakanishi Hiroshi

In postwar history, particularly in the post–Cold War era, it has become a cliché that Japan and the United States share common values and their alliance is built on them. For example, the 1996 Hashimoto-Clinton declaration on security asserted, "The Prime Minister and the President reaffirmed their commitment to the profound common values that guide our national policies: the maintenance of freedom, the pursuit of democracy, and respect for human rights." The official statement by Prime Minister Abe Shinzo and President Barak Obama in April 2014 conveyed the same spirit: "The relationship between the United States of America and Japan is founded on mutual trust, a common vision for a rules-based international order, a shared commitment to upholding democratic values and promoting open markets, and deep cultural and people-to-people ties." Even allowing for diplomatic niceties, these words reflect the fact that the two nations today have very similar views on political, economic, social, and cultural affairs.

Despite the preponderance of overlapping views, there is a significant gap between the two countries, as well as within Japan, in terms of the means to promote these values internationally. On the one hand, the United States, as leader of Western liberal societies in the postwar world, shook off its prewar isolationist impulse and acquired the habit of resorting to enforcement measures, including military might, in order to sustain and sometimes promote liberal values. On the other hand, Japan, particularly before the end of the Cold War and the first Gulf War in 1991, strictly restrained itself from using force beyond its borders. In the aftermath of the first Gulf War, serious debate started on the role of force in Japanese foreign policy, dubbed the "Japan as normal country" debate. Prime Minister Koizumi Junichiro took huge strides toward "normalization" when he sent Self-Defense Force units to the Indian Ocean and the Middle East in the context of the "war on terror" initiated by George W. Bush after the 9/11 attacks. Abe has moved further along this path, and the US government endorses this. As the Joint Statement of the Security Consultative Committee on October 3, 2013, declares: "Japan will continue coordinating closely with the United States to expand its role within the framework of the US-Japan Alliance. Japan is also preparing to establish its National Security Council and to issue its National Security Strategy. In addition, it is re-examining the legal basis for its security, including the matter of exercising its right of collective self-defense, expanding its defense budget, reviewing its National Defense Program Guidelines, strengthening its capability to defend its sovereign territory, and broadening regional contributions... The United States welcomed these efforts and reiterated its commitment to collaborate closely with Japan."

These policy changes, labeled "proactive pacifism" by Abe, however, have aroused fierce debate in Japanese society. After the government announced a new interpretation of the Constitution, including partial exercise of the right of collective self-defense, opinion polls showed that opposition to the new interpretation tends to be larger than its support. Actually, this division on the use of force as a means of diplomacy has been a long-existing feature of Japanese foreign policy debate from the early postwar period. The pacifist Article 9 of the Constitution has been the most powerful rallying symbol of the left, after being proposed by American New Dealers, who sought to realize their progressive ideals in occupied Japan. The US-Japan alliance revised this pacifist spirit, and Japanese officials, including those who revived their political life after the political purge conducted during the occupation, supported the alliance as both a military and an ideological bulwark against communism. When Prime Minister Kishi Nobusuke, the grandfather of Abe, chose to complete the ratification of the revised US-Japan Security Treaty, overcoming an opposition riot while giving up his hope of changing the Constitution, the postwar Japanese consensus was formed. That consensus has a somewhat complicated structure, where the right and left fight unceasingly from opposite ends of the spectrum. Abe is following in his grandfather's footsteps, holding the alliance as the lynchpin of Japanese strength as well as proof of commonality of Japanese worldviews with other advanced democracies.

However, there is skepticism both within and outside Japan that Abe's stance on the war and Japan's colonial past might be stepping out of the postwar consensus. It is true that Abe's association with vocal critics of what they call a "masochistic historical view" and his visit to the Yasukuni Shrine despite requests from abroad, including the Obama administration, suggest sympathy for this line of thought. But in reality, this "correct history" movement is still very much a minority in Japanese society. It sometimes feeds on the public's frustration with China and South Korea, pains they experience from a globalizing and aging society, and the weakening of forces on the left such as *Asahi Shimbun*. Still, it remains a minority and its ideological appeal is largely limited.

The most typical case is Abe's visit to the Yasukuni Shrine in December 2013. Opinion polls on this visit showed almost a 50/50 division, but according to *Asahi Shimbun*'s poll conducted before the visit, about 60 percent supported the prime minister's visit to Yasukuni while less than 20 percent were opposed. The gap apparently widened because of the American expression of disappointment, while China and South Korea issued relatively restrained criticisms. For the young generation in particular, the Yasukuni issue is far more a diplomatic issue of nationalistic rivalry with China and South Korea than an issue of historical remorse and apology. The aforementioned *Asahi* poll shows that 56 percent of those in their twenties know that Yasukuni enshrines the war criminals compared to 84 percent of those in their thirties or above. Younger Japanese tend not to care about the history issue but do focus on nationalistic competition.

Koizumi, more than anybody else, contributed to reframing Yasukuni visits. Ideologically not a conservative and probably even not familiar with past ideological debates over history, Koizumi picked up the Yasukuni issue as a symbolic taboo of Japan's postwar "abnormality" and pledged annual visits to the shrine as part of his "normalizing" agenda. Against the backdrop of the anti-Japanese rioting in China at an Asia Cup soccer game in 2004, President Roh Moo-hyun's campaign to expose alleged colonial collaborators with Japan, heightened attention over the Takeshima (Dokdo) dispute, and renewed textbook controversies, the public took Chinese and South Korean denunciations to be efforts to use the 'history card" against Japan.

Those who propagate through the media and the Internet denials of the atrocities China and South Korea claim Japan committed during its colonial rule and the war have become more conspicuous in Japanese society since the late 1990s. Symbolic of this is the *Sensoron* series by the cartoonist Kobayashi Yoshinori; but a closer look at the series (three volumes in total) and Kobayashi's track record shows the weakness of this historical revisionism as an ideological movement. First, associating with politics in the course of drug-induced HIV lawsuits, Kobayashi initially had no strong attachment toward the right wing. Then, disillusioned with the left wing and stimulated by the historical textbook rewriting movement, he wrote the first volume of *Sensoron* in 1998, most of whose content was borrowed from old rightist Asianist ideology. But the history textbook movement soon hit a wall and broke up into pro-American and anti-American groups. Kobayashi was, for a while, in the anti-American group, and the second *Sensoron* in 2001 was partially sympathetic to al-Qaeda's 9/11 attack. But in the third *Sensoron* in 2003, even though critical of the war in Iraq, he became much less vocally anti-American, and the main theme shifted to emotional sympathy with the *Tokko* (kamikaze) fighters' self-sacrifice. Later, Kobayashi distanced himself from the historical revisionist movement, expressing approval for a female imperial succession, which is a minority view among today's rightists.

In the meantime, the history textbook movement itself lost momentum, divided into anti-Chinese, anti-Korean, and more broadly nationalist groups. This transformation was caused by many factors from personal animosities and political differences to media strategy, but the absence of any unifying ideology was the key. Anti-Western Asianist ideology survived until the 1970s, but the shocking suicide of the novelist Mishima Yukio in 1970 and the deaths of the foremost proponents of this ideology, such as Hayashi Fusao and Takeuchi Yoshimi, in the late 1970s virtually marked its end. The later rightist movement borrows from its rhetoric, but given the deep Westernization of Japanese society, the rhetoric always turns out to be somewhat restrained.

It is an interesting coincidence that history conflicts with China and South Korea essentially arose from the 1970s when Asianism was ebbing and the psychological settlement with the West was institutionalized. The Nanjing massacre and the "comfort women" issue, for example, were known from the early postwar era, but they came to receive serious attention only when leftist journalists and writers such as Honda Katsuichi and Senda Kakoh took them up in the early 1970s. It may well have been intended to buttress the weakening leftists, but the textbook controversy in 1982 with China and South Korea put the history issue on a different plane of international politics. The criticism of Prime Minister Nakasone Yasuhiro's visit to the Yasukuni Shrine in 1985 by China added a diplomatic aspect to the history issue. In this sense, the current history debate in Japan is relatively new, which can be compared more to the *Historikerstreit* in the late 1980s, which was about comparing Nazi and communist atrocities than to the postwar reconciliation between West Germany and France.

This renewed interest in past Japanese deeds resonated with the global trend of historical revelations and apologies to the victims of past wars and discrimination from the late 1980s. A case in point was the great attention paid in the United States to *The Rape of Nanking* published by Iris Chang in 1997, though its historical precision was questioned. The report submitted to the UN Human Rights Council in 1996 on "wartime sexual slavery" as an addendum to the report on violence against woman by Radhika Coomaraswamy was another example.

This revival of interest in the wartime and colonial past took Japan off guard. Postwar Japan took the stance that war-related apologies and reparations had been taken care of by the peace treaties and their equivalents and that the reformed Japan was to have future-oriented friendships with former enemies. The Japanese government held to the principle that legal reparations were completed by those treaties. Still, the government in the early 1990s managed to issue some statements of remorse and apology, notably the Kono statement in 1993 on the "comfort women" and the Murayama statement in 1995 on Japanese wartime conduct. However, these official moves resulted in a backlash in society, as with the new history textbook movement. The movement was boosted by North Korean missile launches and Kim Jong-il's revelations to Koizumi in 2002 of past abductions of Japanese, which discredited leftists who had been sympathetic to North Korean claims including North-related Korean organizations.

The tendency to see the history issue with China and South Korea in a nationalist light fueled extremism. *The Citizens Group against the Privileges of the Koreans in Japan* (*Zaitokukai*), who demonstrate and chant violent slander in Korean communities, has become conspicuous. Anonymous Internet messages are filled with vile language. However, while they should not be ignored, these unfortunate tendencies exist in other Western societies and in Japan they are small in number and receive strong criticism. The vast majority of Japanese are admittedly frustrated with the nationalist strife with China and South Korea, but they generally admit Japan's guilt for the war and past colonial conduct.

Of course, nationalism is something East Asia, including Japan, needs to overcome. Japan is struggling with this. In the United States or Europe, new revelations of the past, an objective search for historical truth, transnational cooperation among civil societies, and political settlements with former adversaries go hand in hand. In East Asia, such conditions are hard to come by. The Japanese government endorsed an historians' dialogue with South Korea and China and helped establish the Women's Fund to send compensation to former "comfort women" abroad, but, despite huge efforts by the people involved, the result has not been impressive.

Given the undemocratic nature of the Chinese regime, Chinese historians are understandably hesitant to differ from the official historical line. In Japan, most historians admit the occurrence of a mass killing at Nanjing by the Japanese army in 1937, but they do not support the number of victims (totaling 300,000) officially claimed by the Chinese government. No doubt the fact of the atrocity itself is more important than the number, but the way the Chinese official line blocks objective and independent dialogue weakens the legitimacy of attempts to find common ground and feeds extreme criticism. South Korea's predominantly moralistic denunciation of Japanese colonial rule has made the dialogue mediated by sincere liberals both in Japan and South Korea extremely painful and costly. The Women's Fund made some headway with former "comfort women" in Southeast Asia and with European detainees, but in South Korea its activities were blocked by Korean activists because the attempt left ambivalent Japanese state responsibility. The historians' dialogue succeeded in publishing reports, but the gap between positivist Japanese historians who put truth finding first and moralistic Korean historians who claim that moral judgments cannot be separated from historical truth tends to exhaust both sides.

With the seventieth anniversary of the end of the war approaching next year, the history issue will become an even tougher challenge for Japan. Given the widespread nationalism in Japan, China, and South Korea, no single deed or statement by leaders is likely to change the situation dramatically. Even so, Japan is trying to face its historical past as the Western liberal societies have done, and put nationalism back into a box. By understanding the Japanese domestic debate, the outside world, including the United States, has much to offer to assist Japanese efforts on this issue. The politics of history in Japan have problems to be sure, but they, in no way, indicate that Japan is moving away from the universal liberal values fundamental to the current international order.

Part 3: A South Korean Perspective

Bong Youngshik Daniel and Hyun Daesong

Do the United States and Japan share the same values and overall view of history? Answering this question and examining how South Koreans think about it is to place the question in the context of the US-Japan security alliance. There are two reasons why this exercise is useful. First, military alliance is a good barometer to measure the strength of mutual trust between countries based upon common values and worldviews. Forging a military alliance is not only determined by the purpose of deterring the projected military threat from a common adversary. Military alliance is also an institution that nations create to protect political ideology and key principles they deem indispensable for maintaining civilized orders. For instance, as political scientist Tony Smith concludes in his book *America's Mission*, World War II marked the defeat—the former immediate and the latter after four decades—of fascism and communism, the two totalitarian rivals of liberal democracy, as viable forms of political organization. It was not just a military victory by the Allied Powers.

Second, the question is closely related to the issue of upgrading the US-Japan security relationship as a value-based, future-oriented alliance into the twenty-first century. In the Joint Statement of the Security Consultative Committee issued after the 2 + 2 meeting on October 3, 2013, both governments declared that they "set forth a strategic vision, reflecting our shared values of democracy, the rule of law, free and open markets, and respect for human rights." The 2013 Joint Statement also emphasized that "promoting deeper security cooperation with other regional partners to advance shared objectives and values" is one of the objectives for the revision of the 1997 Guidelines for US-Japan Defense Cooperation. Prime Minister Abe Shinzo claimed at the joint press conference with President Barak Obama on April 24, 2014, that "between Japan and the United States, we share values such as freedom, democracy, human rights and rule of law. We are global partners."

Enhancing the bilateral security partnership as a value-based military alliance is an ambitious goal for both sides. Under international anarchy, where there is no central authority above sovereign states to enforce promises between states, it might be regarded as a rarity that states tie their national security to pursuit of shared values. To realists, it is a futile and dangerous practice. Justifying your security alliance with values and principles is only useful as nice diplomatic rhetoric or a code word. Yet, others consider that the transformation of the US-Japan alliance, if successful, would serve as a beginning of the institutionalization of the Asia-Pacific security community based upon common values and a collective identity, something akin to NATO. As Thomas Risse-Kappen explains, compared with NATO and the EU, the US-Japanese security relationship appears to have remained weak with regards to the level of mutual identification with the same civilization, although they succeeded in establishing norms of consultation and policy coordination similar to those of NATO. Walter Lippmann explained that the members of the "Atlantic community" are "natural allies of the United States." The day when the US-Japan military alliance constitutes a pluralistic security community similar to NATO, we will be able to conclude that a sense of shared values between the two countries and their peoples is indeed strong and fundamental.

South Korea has closely followed the debates on the United States and Japan sharing common values and historical view. The main reason for this interest is that, although it is in essence an issue between the United States and Japan, it has direct and significant impact on Korea's foreign relations with the two countries and its national identity politics. South Korea has a keen interest in this issue from the alliance perspective. Broadly speaking, it assesses the status of the US-ROK military alliance by two standards: functional and comparative. As for functional aspects, the South Korean government and public assess the value of the alliance in terms of its contribution to national security, especially for maintaining a sufficient and reliable deterrence and defense capability against military threats from North Korea. At the same time, South Koreans tend to use the US-Japan alliance as a measuring stick for US "fairness" toward South Korea as its ally. The way the United States and Japan define the core missions and nature of their bilateral security alliance affects the way South Koreans expect the United States to define those of the US-ROK security alliance.

To many South Koreans, the US-ROK security partnership must be as qualified as the US-Japan security alliance is to become a global partnership based upon common values and historical views. Like Japan and the United States, South Korea and the United States have taken steady steps to elevate the status of their alliance to a value-based alliance. The Joint Vision for the Alliance of the United States of America and the Republic of Korea announced on June 16, 2009, stipulated the commitment of both governments to "build a comprehensive strategic alliance of bilateral, regional and global scope, based on common values and mutual trust." The 2009 Joint Statement even tied the mission of the alliance to Korea's unification based upon shared values between the allies. It defines the purpose of the alliance as "establishing a durable peace on the Peninsula and leading to peaceful reunification on the principles of free democracy and a market economy." Such strategic vision is reiterated and articulated in the 2013 Joint Declaration in Commemoration of the Sixtieth Anniversary of the Alliance between the Republic of Korea and the United States of America, in which the two declared that the alliance "has evolved into a comprehensive strategic alliance with deep cooperation extending beyond security to also encompass the political, economic, cultural, and people-to-people realms. The freedom, friendship, and shared prosperity we enjoy today rest upon our shared values of liberty, democracy, and a market economy." The 2013 Joint Declaration also affirmed that it is the basis of the Joint Vision that Korean unification should be achieved peacefully and "based upon the principles of denuclearization, democracy and a free market economy."

Some may suggest that finding out whether the alliance can be a genuine value-based security partnership in fulfillment of the official strategic visions is impossible until North Korea's military threat disappears. Only then will we be able to find out if the alliance is truly based upon mutual identity and common values, as in the case of NATO remaining robust even after the disappearance of the Soviet Union. To this suggestion, one can point out that Japan is not free from the same problem either. One can only imagine what the main strategic vision underpinnings of the US-Japan alliance might have been had there not been the rising military power of China. The discourse of a value-based US-Japan alliance would not have been so strongly popularized in the absence of the China threat.

Another reason for South Korea to pay close attention to the issue of the US-Japan security alliance as a value-based alliance is that it deeply touches upon Japan's remembrance of the colonial occupation and the Pacific War and its attempts to achieve its identity as a "normal country." This is not just a domestic issue for Japan; for the Korean people, it is very much an issue related to the sovereignty and national character of Korea. As illustrated in the Preamble of the Constitution of the Republic of Korea, the national identity is inseparable from its struggles against Japanese militarism and colonial occupation, upheld by "the cause of the Provisional Republic of Korea Government born of the March First Independence Movement of 1919." This is why how the United States positions itself in debating and understanding Japan's war responsibilities affects South Korea's willingness to support the US-led effort in forging a virtual alliance between the United States, South Korea, and Japan as like-minded countries.

Political and intellectual leaders in Japan, both from conservative and liberal orientations, concede that Japan needs to become a "normal country," which it has not been since its defeat in the Pacific War. The search for normalcy as a sovereign country has been largely dominated by the conservatives. The LDP used its fiftieth anniversary to iden-tify 2005 as the first year Japan would start revising its Constitution, focusing on territorial rights and history reeducation that emphasized nationalism and tradition. It declared that 2005 would be a year of "complete self-reassessment of the postwar system." The reassessment prompted a shift in government policies and more aggressive actions, leading to an explosion of diplomatic rows and conflicts with Japan's neighboring countries.

The conservative right-wing party has formed committees to serve as core actors on both domestic and foreign affairs: "History Review Committee," "Politicians' Meeting to Pay Homage to the Yasukuni Shrine," "Junior Politicians' Meeting for Those Who Support Visiting the Yasukuni Shrine to Pray for Peace and Promote National Interest," "Union to Protect Japan's Territory," "Group Meeting to Promote Japan and History Education," "The People of Japan," and "Association of Politicians Who Pursue Value Diplomacy." They are deeply involved in history, territorial, and security related issues, directly responding to diplomatic conflicts and having a significant influence on the policy decision-making process, which strengthens their conservative base.

The landslide victory by Abe's LDP in the fall of 2012 elections and the party's win in the Upper House in the summer of 2013 consolidated Japan's right-wing power. Opposition parties basically fell apart; the so-called progressive-liberals were aging and weak, and the main opposition party split. Even civil society groups that had criticized the movement toward ultraconservatism had weakened. Under such circumstances, Japan's right-wing politicians took center stage and carried out acts that viewers in South Korea and many others perceived to be regressing to Japan's pre-1945 past.

Such efforts by the conservatives for the past decade to ground the quest for a "normal Japan" in reconstructing memory of the 1930s–1940s have complicated Japan's relations with the United States. These efforts challenge American values and views of modern history, even as the Abe administration has deepened Japan's dependency upon the United States as the essential security partner. Just as we have witnessed for the past two years that Japan has actively supported the US rebalancing strategy toward Asia, its top leaders have been making remarks indicating that a great civilizational divide exists between the two countries on political and historical issues.

For instance, Prime Minister Abe held a summit with President Obama only two months after his inauguration, fulfilling his promise to "restore" the US-Japan security tie that had been severely damaged during the DJP government as his top priority. But only two months later in April, Abe made a controversial remark at an Upper House Budget Committee session that embraced a revisionist view of Japanese history, saying that what constitutes aggression during World War II has not been settled. The controversy left the impression that the man Obama had embraced as his counterpart in security and in a values agenda seemed to believe that Japan basically was forced into fighting World War II and the aggression leading up to it, including the Pearl Harbor attack, because it was motivated by legitimate concerns over Western colonial policies in Asia.

Yoshida Yutaka proposes "double standards" as the key words in understanding the way in which the Japanese government and conservatives interpret Japan's war responsibilities.[1] He argues that, at the international level, the Japanese government made a decision to accept war responsibilities at a minimal level by accepting the outcome of the Tokyo Tribunal, as affirmed in Article 11 of the San Francisco Peace Treaty. In return, Japan acquired the status of US ally. Domestically, the government virtually denied the war responsibilities by holding only the military culpable, but leaving the citizens as victims. Such a dual approach to the issue of war responsibilities encouraged the Japanese to identify themselves as victims deceived by the military rather than perpetrators of atrocities upon people and nations in the Asia Pacific. The United States helped to minimize Japan's sense of responsibility by maintaining very generous occupational policies in order to build an anticommunist security system during the Cold War. The sense of victimhood underlying the collective psychology of the Japanese people accounts for the popularity of Kobayashi Yoshinori's *Sensoron*, which espouses that the Pacific War was a just war for the liberation of Asia from Western colonial powers.

Shirai Satoshi submits that postwar Japan has been locked in the state of *eizoku haisen* (permanent defeat),[2] a kind of vicious cycle in which Japan is perpetually dependent upon the United States in order to remain free from facing its war responsibilities in any fundamental way. As long as it remains dependent upon the United States for national security, Japan can continue to deny accountability for its wartime crimes. Although this security dependency allowed Japan to achieve postwar prosperity and peace, it deepened its isolation from neighboring countries. In return, isolation in the region provided additional motivation for it to more strongly depend on the United States. This cycle of isolation and dependency produced a grotesque system in which the foundational identity of nationalism was supported externally and patriotism and conservatism became pro-American, rather than critical of the US influence.

The Abe government appears to have inherited this system of *eizoku haisen*. On the one hand, it has aggressively supported the US pivot to Asia with a number of important government restructuring measures and bold revisions of the law on national security, including the cabinet decision that allows Japan to exercise the right to collective self-defense. On the other hand, the government has committed a series of controversial actions on historical issues, including Abe's visit to the Yasukuni Shrine and the decision to review the process of making the 1993 Kono statement, which admitted Japan's responsibility for the comfort women. This pattern is ironic in that it made important contributions to the successful implementation of rebalancing to Asia in the short term, but it undermined the long-term prospects of the rebalancing. The historical revisionism of the Abe administration escalated hostility between China and Japan and thereby raised the cost of US military engagement in East Asia and prevented South Korea—another crucial US partner for implementing the rebalancing strategy—from fully joining the US-led regional security initiatives. It also collides with core cultural values for which America stands, undermining the ideational message that Washington is seeking to convey. The US-Japan security alliance cannot be transformed into a genuinely value-based alliance as long as Japan fails to complete its quest for the state of normalcy as a sovereign state.

Notes

1. Steffi Richter, "The 'Tokyo Trial View of History' and Its Revision in Japan/East Asia," in Gotelind Mueller, ed., *Designing History in East Asian Textbooks: Identity Politics and Transnational Aspirartions* (Routledge: Abingdon, Oxon, 2011).
2. Shirai Satoshi, *Eizoku haisenron: Sengoku Nihon no kakushin* (Tokyo: Ota shuppan, 2013).

PART V

Japanese and Korean National Identity

CHAPTER 16

National Identities, Historical Memories, and Reconciliation in Northeast Asia

Gi-Wook Shin

In her speech to a joint session of Congress in May 2013, Park Geun-hye contended, "Asia suffers from what I call 'the Asian paradox,' the disconnect between growing economic interdependence on the one hand, and backward political, security cooperation on the other." This is, she noted, because "differences stemming from history are widening" and "how we manage this paradox" will determine the configuration of a new order in Asia.[1] Other leaders of the region would agree with her assessment, though they might differ on how to manage the paradox. While Northeast Asia has witnessed growth in regional interactions over the past two decades, especially in the spheres of culture and economy, wounds from past wrongs, committed during colonialism and war, have not yet fully healed. The question of history has become a highly contentious diplomatic issue and a centerpiece in national identity, crowding out other dimensions and complicating bilateral relations as well as US strategic calculations.

One year into office, the new leaders of China and Korea still refuse to hold a bilateral summit with their Japanese counterpart, outraged by remarks made by Abe Shinzo and his cabinet members on historical and territorial issues. His visit to the Yasukuni Shrine in December 2013 only aggravated tensions. Park has insisted that "relations with Japan are not a matter of summits; this is a problem that needs to be solved with the Korean people."[2] Xi Jinping likewise urged that "Japan should face up to history and keep an eye on the future, correctly deal with such sensitive issues as the Diaoyu Islands and questions of history, and seek a way to properly manage differences and address problems."[3] Japan, in contrast, continues to complain of "apology fatigue." Foreign Minister Kishida Fumio told John Kerry that Japan has put forth maximum effort to resolve regional history issues in a sincere manner.[4] Abe also maintains that the definition of aggression is vague academically and internationally, and "as there is not yet an academic consensus on the matter of defining the meaning of invasion, I will not get involved in that task."[5] So much attention is being put on these historical issues that Japan and its two neighbors are increasingly viewing each other through this prism.

A 2013 survey conducted by *Asia Today* and *Realmeter* in Korea revealed that 66.1 percent of respondents feel that Japan has not apologized to Korea about its past wrongdoings in an appropriate manner; 30.9 percent that Japan has apologized, but that the apologies were lacking in sincerity; and only 1.6 percent that Japan has apologized sufficiently.[6] Furthermore, 91.6 percent responded that Japan must reapologize—in a sincere manner—for its role in regional history, a figure that reveals the high dissatisfaction about the way Japan's apologies have been delivered. In contrast, a 2013 survey jointly conducted by *Seoul Shinmun* and *Tokyo Shimbun* shows that 63.4 percent of Japanese find Korea's continued demands for apology incomprehensible and react negatively with "apology fatigue,"[7] indicating a widening perception gap between the two countries. History issues overwhelm other perspectives on the national identity gap. They raise questions also about the intertwining identity gaps of multiple countries. If the main stress here is the ROK-Japan gap, efforts to understand the fundamental character of such gaps and to propose steps to narrow them in a lasting manner lead to consideration of the China-Japan gap and, with further investigation, of the Japan-US gap.

Historical Memory, National Identity, and Regional Relations

No one can dispute that nationalism is a key factor in provoking tensions between Japan and Korea[8]; however, this is such a broad concept that one must consider specifically what triggers "nationalist" sentiments in both Japan and Korea. Here I focus on the role of historical memory in shaping national identity, which, in turn, affects regional relations. Questions about history touch on the most sensitive issues of national identity, including the national myths that play a powerful role to this day.

Whether it be Japanese atrocities in Korea and China or the US decision to drop atomic bombs on Japan, no nation is immune to the charge that it has formed a less-than-complete view of the past. All share a reluctance to fully confront the complexity of their past actions and blame others for their historical fate. Divided historical memories or a gap in identities separates nations, resulting in distinct, often contradictory, perceptions deeply embedded in the public consciousness transmitted to succeeding generations formally by education and informally through the arts, popular culture, and mass media.[9] They can become central to a population's sense of its shared identity, superseding all other identity themes.

Why have these nations developed different, distinct, and incomplete historical memories? One common answer would be that Japan is an aggressor and Korea/China is a victim. While this is historically true, a closer examination reveals that the aggressor-victim dichotomy is too simple to explain the complexities of modern history and collective memory in Northeast Asia. Historical memories and reconciliation are rooted not only in colonial and Pacific war injustices but also in much deeper and more complex historical, cultural, and political relations. We need to understand the crucial imbalance in each country's formation of historical memory that leads to the gap in national identities. Certain past experiences symbolize how the identities of two nations differ, shaping how other eras in bilateral history are viewed and how cultural or political divisions are seen.

Different factors figure in the formation of historical memory in each of the involved countries in the region. For example, for China and Korea, Japanese acts of aggression—such as the Nanjing massacre or forced labor and sexual slavery—constitute the most crucial element in their wartime memories, while for Japan, memories related to US actions, such as the fire-bombings of Japanese cities or the nuclear attacks on Hiroshima and Nagasaki, are the most crucial in their memories of the war. This imbalance in focus is also noticeable in each country's history textbooks. In Japanese ones, only 4 percent of the coverage of Japan's modern history (1868–1945) is devoted to Korea, while the United States is the main player.[10] In contrast, in Korean ones, Japan occupies a quarter of the coverage of its modern history.[11] Korea and China are a less significant element in Japan's wartime memories, while Japan figures most prominently in their memories.

This imbalance in focus helps to explain the formation of historical memory in each nation. Japanese focus on US actions in war memory both reflects and explains Japan's victim identity and reluctance to fully engage Asian neighbors regarding colonialism and issues of wartime aggression. Unlike Germany, postwar Japan developed a mythology of victimhood in which many innocent civilians were sacrificed as a result of the massive and destructive atomic and fire bombings of its cities. Many Japanese charge that the United States did not address its own "crimes against humanity." Memory based on victimhood then pardoned the Japanese of their guilt, fostered an already ubiquitous sense of self-pity, and impeded the search for historical truth. Victim consciousness provided fertile soil for the growth of postwar neonationalism that denied Japan's responsibility for wartime atrocities. This imbalance in historical memory formation produces misperceptions and misgivings among the parties, hindering reconciliation.

Historical disputes did not emerge as a key foreign policy issue until the 1980s. During the Cold War years, they were put aside, as in the normalization of relations between Japan and South Korea in 1965 and between China and Japan in 1972. Japanese wartime atrocities and compensation were not to be raised, as governments focused on economic development, which was heavily dependent on Japanese capital and investments. The need to appease Japanese sensibilities (and US interests) meant that it was considered unwise or impractical to recall the horrors of the war or the colonial period in official discourse. The history question began to surface in response to the 1982 controversy, in which Japanese authorities attempted to distort the connotations of some words in its 16 types of history textbooks, for example, removing the term "invasion" and replacing it with "advance," or replacing "exploitation" with "transfer." Up to this point, textbooks were almost purely a domestic issue, mainly within Japan. Yet with the emergence of civil society in the 1980s, issues of historical injustice were no longer monopolized or controlled by governments. Following South Korea's democratization and the "comfort women" issue's resurfacing as part of a "transnational memory with social, legal, and moral consequences that transcend national or cultural borders," the emergence of a new "community of memory" in Korea recast the war in international rather than purely domestic terms. These changes in domestic and international environments brought the history question to the forefront of Northeast Asian relations as the Cold War ended.

National identity is not necessarily obsessed with the past. In the past decade there were times when attention focused on other dimensions, such as the East Asian community and shared universal values. Even if the United States is inclined to set the past aside, East Asian states are not, and they bring previous US behavior into their identity concerns. In this chapter the identity gaps between Japan and China and Korea, respectively, are linked to identity gaps with the United States, as we consider ways to narrow the sharp divides. A framework of interconnected identity gaps is proposed, leading to suggestions to deal simultaneously with multiple gaps as a way to overcome barriers to narrowing each one.

The Complexity of the History Question

The growing complexity of the situation in Northeast Asia does not bode well for the future. First, tensions over history and historically rooted claims to territories are not occurring simply among adversaries. Conflict exists also between two key allies of the United States, Japan and South Korea, taking a turn for the worse in recent years, making the United States increasingly concerned and frustrated: "This has now emerged as the biggest strategic challenge to American interests in Asia,"[12] said Kurt Campbell, who served as assistant secretary of state for East Asian and Pacific affairs during President Obama's first term. Daniel R. Russel, Campbell's successor, has also voiced his concern that "the headwind created by these tensions over history raises the political cost of Japan-Korea cooperation that should be a given."[13]

Some experts approach the history question from a realist perspective. Victor Cha, for instance, poses what he calls the *quasi-alliance paradox*—that the two unallied states will better align and cooperate when there exist concerns of abandonment and entrapment by a common third-party ally, that is, the United States—and argues for a gradual and calculated disengagement from the region by the United States.[14] However, this approach cannot adequately explain the Japanese-Korean antagonism because the United States is not just a common great-power protector, but an integral part of the problem itself.

Second, tensions over the past have not improved with the passage of time. Those who did not experience war years are seemingly more critical of the history issues than those who did. According to a 2013 survey of Koreans, 42.8 percent of respondents in their twenties—who have no firsthand experience of Japan's past wrongdoings and perhaps are more familiar with Japanese cultural influence—identified the question of history as the most critical issue to be resolved, while only 34.6 percent of those in their sixties who lived through the period in question viewed it as the most pressing problem in relations.[15] This points to the importance of political leadership and education, official and unofficial (such as media or cultural and educational exchanges). Not only are identity gaps best addressed at many levels, they lend themselves to an inclusive identity approach with multiple "significant others" entering the picture in search of a comprehensive strategy.

"The Asian paradox" clearly attests to the power of historical memories and national identity in shaping regional relations in Northeast Asia. Realist scholars argue that Japan and other governments are behaving "rationally" to maximize their current interests through manipulation of public sentiment over the history issues. Alexis Dudden counters that this line of reasoning treats history as occurrences in "homogenous empty time," where past events are measured as equal to events occurring in the present; from the past, moments, heroes, and villains are selectively chosen for the present. She contends that this overlooks history—as if it were yet another factor of the present, like a trade imbalance or background music—and limits our understanding of why the historical problem weighs so heavily on Japanese-Korean relations, among others in the region.[16]

Limits of Past Efforts at Reconciliation

Divided historical memories have hindered efforts at reconciliation, even as there has been widespread recognition of the need for final resolution of historical injustices. In fact, all involved countries have, to a certain extent, sought to achieve this goal through various means—apology politics, litigation, joint history writing, regional exchanges, and so on. Here, I discuss how different foci in historical memory formation have hindered reconciliation, using the examples of apology and joint history writing.

Apology

Apology diplomacy has been a major tactic in the reconciliation process. Since 1984, Japanese heads of state or government have issued a number of direct apologies to China and Korea. Until the 1990s, the key terms used were "regret" or "remorse," which did not necessarily signify an apology, a term that first appeared in 1992, when Miyazawa Kiichi meeting with Roh Tae-woo stated, "I, as prime minister, would like to once again express heartfelt remorse and offer an apology to the people of your nation."[17] Despite Japan's repeated use of the term "apology" from that point on, its neighbors have continued to respond with skepticism. As Caroline Rose points out, Japan's various efforts to "apologize" have not been backed up by actions to "reinforce the apologies"; instead, they are often coupled with ambiguous wording and counterproductive statements and behavior on the part of the government. In fact, throughout the 1990s, Japanese political elites vacillated between formal apologies and frequent statements glorifying their colonial rule.[18] Repeated statements by right-wing officials, such as Abe, denying war crimes—"there is no evidence that comfort women were coerced through violence or coercion by military or government authorities"—aroused severe criticism.[19]

Along with Abe's attempt to revisit the Kono statement and his recent visit to the Yasukuni Shrine, the outpouring of ultra-nationalist Japanese books, films, and magazines raising doubts about the veracity of their past aggression has led neighbors to question the sincerity of the Japanese apology.[20] People in Korea and China have gradually realized that the formal ritual of apology is but one element in the politics of remembrance with questionable utility as a means of furthering historical reconciliation. According to Dudden, politics surrounding state-issued apologies have largely negated the putative intent of apologizing and, if anything, have set Japan back in terms of actually reconciling with its neighbors. She sees "apology failure" (apologizing for the past as a means to capitalize on it in the future)—not failure to apologize—and asserts that the general effect has only "perpetuated a disastrous policy failure."[21] Decades of this have created a deadlock in the East Asian reconciliation process that will not be broken until Japan addresses its "identity mythmaking."

Joint History Writing

As history education plays such a powerful role in the formation of collective memory and national identity, collaborative history writing has been another approach toward narrowing the gap in views. The frequent clashes over history textbooks in Northeast Asia—in 1982, 2002, 2005, and 2011, for instance—demonstrate that history is not simply about the past. Government oversight makes textbooks a natural and legitimate subject for debate, and it is no coincidence that textbooks have become a nexus for tension in the region.[22] One approach has been to form official and unofficial committees to produce jointly written accounts of history.

The first official attempt to deal jointly with history textbooks occurred in October 2001 when Koizumi and Kim Dae-jung established the Japan-ROK Joint History Research Committee, a state-sponsored effort toward placing a reconciled view of the past in a new regional history framework. The committee, while not entirely a failure, has yet to attain the success envisioned. Although it adopted the UNESCO model of writing a "parallel history" in May 2005, the two sides failed to come to a consensus on what should be incorporated into the textbooks, disagreeing over how to interpret Japan's colonial rule, including its role (or lack thereof) in Korea's modernization.[23] Following Koizumi's visits to the Yasukuni Shrine, the work of the joint committee was put on hold until October 2006 when Abe and Roh Moo-hyun relaunched its efforts. In April 2007 a new subgroup—in addition to existing groups studying ancient, medieval, and contemporary history—formed to study history textbooks to try to narrow differences between the textbooks of the two nations. Another report was released in 2010 but still failed to reach a consensus on Japan's 1910–1945 colonial rule, notably on recruitment of Korean laborers and "comfort women" as well as on drafting Koreans into the Japanese military.[24]

Japan and China launched a similar effort as part of the thaw that followed the leadership transition from Koizumi to Abe. Modeled after the Japan-ROK format, the Joint History Research Committee of 20 leading historians from both countries aimed to write parallel histories. They agreed to conduct a joint study and produce an account of 2,000 years of Sino-Japanese interaction by 2008. Not surprisingly, the Japanese wanted to focus on the postwar era, while the Chinese were more interested in taking inventory of the colonial and wartime periods.[25] Each side was to separately write its own version of bilateral history texts and exchange written comments on controversial issues. They agreed on a list of major historical events to be discussed, including the Nanjing massacre and Japan's Twenty-One Demands.[26] The final report was made public in January 2010, after one and half years' delay,[27] only to show that the two sides could not resolve differences on controversial modern events including the 1937 Nanjing massacre.[28]

Writing a shared history has offered important lessons for the nations involved. It seems impossible to arrive at a common rendition of events, particularly regarding the most controversial aspects of history. A shared regional history is not feasible politically. Northeast Asian governments have considerable influence in textbook production, and these books have become a nexus of significant tension in the region. The experience of the past two decades underlines how profoundly historical writing—especially the writing of history texts—is affected by nationalist politics.

The US Role

Even when the governments of Japan, China, and South Korea have tried to stop being held hostage to history, public perceptions have not always followed. This tension is increasingly unfavorable to US strategic interests. According to a May 2013 report by the Congressional Research Service, "Comments and actions on controversial historical issues by Prime Minister Abe and his Cabinet have raised concern that Tokyo could upset regional relations in ways that hurt U.S. interests."[29] There has been ongoing debate in US academic and policymaking circles about the role that the United States might play in helping to resolve historical disputes and achieve reconciliation. A predominant view is that this is primarily a matter for Asians and better left to historians. Many fear that the United States could be forced to choose between its key allies in the region.[30] The State Department has consistently taken the position that the San Francisco Treaty protected Japan from demands for compensation from victim nations.

An emerging view contends that the United States can hardly afford to stand outside these disputes, particularly as it was a key player in their formation. Referring to the recent dispute over the naming of Dokdo/Takeshima, Dudden aptly points out that it has "brought us back to 1952, when America's occupation of Japan ended, and the United States determined who owned what in East Asia and the Pacific." She asserts, "Washington must not overlook its place in the problem now."[31] Gilbert Rozman argues that US efforts need to be directed more toward narrowing the historical divide in East Asia, while acting as the impetus for increased mutual understanding. He writes that "benign neglect of Japanese nationalism threatens to unravel the spirit of reconciliation in East Asia."[32] Daniel Sneider contends that the United States must "abandon its position of neutrality on wartime history issues, as it is not really a neutral party, and step forward" and play the same central role it played during the Clinton administration in the complex negotiations with governments and citizen groups that led to the formation of the German Fund for the Future—a joint project of the German government and the German private corporations that used forced labor during World War II to compensate survivors in almost one hundred countries.[33]

As the history conflict has become an increasingly vexing problem, the United States has tried various ways "to create the conditions where both sides can find a solution to this."[34] However, any attempt, as a "neutral" player, to broker a deal between Japan and South Korea will prove to be insufficient. It can hardly afford to act as a mere bystander; it was the US-led Tokyo War Crimes Tribunal and the San Francisco Peace Treaty that largely failed—albeit not always intentionally—to address the sufferings and grievances of Asian victims of Japanese aggression.

Unlike the Nuremberg court, only 3 of 11 judges at the Tokyo trial represented Asian countries, and there was no representative from Korea. The US-led tribunal failed to appreciate the massive suffering of Chinese and Koreans at the hands of Japanese invaders and colonizers and the need to dry up the deep well of anger left behind. Instead, the tribunal focused on the Japanese actions that had most directly affected the Western allies—the surprise attack on Pearl Harbor and the mistreatment of Allied prisoners of war. The failure to address the issue of Emperor Hirohito's war responsibility greatly shaped the ways in which the Japanese would remember the war years and later approach reconciliation issues with their Asian neighbors. Encouraged by the American decision, the Japanese elite promoted a myth of the emperor's innocence that only strengthened Japanese victim consciousness and impeded the search for historical truth. A report noted that "the absolution of the Emperor left the country without anyone to blame" and "provided fertile soil for the growth of a postwar neo-nationalism."[35] Acknowledging that it paid scant attention to Asian issues during the Tokyo Tribunal, the United States could set a new tone for recognizing the linkages in historical memories.

As Japan's importance as a bulwark against communism increased in the Cold War, the United States sought to quickly put aside issues of historical responsibility. It did not press Japan to reconcile with its neighbors as it had Germany. The San Francisco Peace Treaty of 1951 formally ended the war, settling Japan's obligations to pay reparations for its wartime acts. But neither China nor Korea was a signatory, and Japan's responsibility toward those nations was left unresolved. Drafted at the height of the Cold War, it denied countries legal means to obtain redress. In 1965, under heavy pressure from a United States anxious to solidify its Cold War security alliance system, the ROK agreed to normalize relations with Japan despite strong domestic protests. Issues such as disputed territories and Japan's colonial rule were again overlooked. Unlike the push for Franco-German reconciliation to establish a multilateral security system in Western Europe, in Northeast Asia the United States established a bilateral "hub and spoke" alliance system with Japan and the ROK and did not press for historical reconciliation between the two US allies.[36] "Normalization" occurred at the governmental level but without addressing popular demands for redressing historical injustices. As one former US senior diplomat points out, "for American policymakers, strategic considerations have consistently trumped issues of equity in historic disputes involving Japan since Word War II."[37]

Looking back, the US approach made sense from a geopolitical perspective, but not in terms of historical reconciliation.[38] It played a crucial role in the rise of Northeast Asia as a powerhouse but sowed the seeds for current tensions over the past. This gave Japan the leeway to escape the responsibilities and penalties of a war that left unhealed scars in China and Korea, while also deliberately leaving the status of Dokdo/Takeshima in a gray area. Taking a self-critical approach by acknowledging, if not necessarily apologizing for, the mishandlings of history issues in Northeast Asia, will endow it with moral power to facilitate a larger process of historical reconciliation in the region as a responsible stakeholder. Focusing national identity on linkages between historical memory gaps in three pairs of states and universal values, which can be reinforced in all of the states, the United States can play a leadership role that transforms the way recent disputes over symbols of history are seen.

Looking Forward

Despite significantly increased regional interactions, the importance of the past has not diminished. On the contrary, it has become even more contentious as nations vie for regional leadership, and there exists an urgent need to find a Northeast Asian strategy of reconciliation. Sustainable reconciliation is inherently a complex process that involves a multitude of actors from the state, civil society, and international organizations. The United States can play a facilitating role. In the Cold War era it put priority on geopolitical concerns over historical reconciliation, but it is becoming increasingly difficult to pursue this strategy. As recent tensions between its two allies show, the history question may no longer be separated from the security question.

Efforts to bridge gaps in understanding history and forge a balanced perception should be underpinned not only by relevant governments but more importantly by civil society. Joint efforts such as group discussions by young people from China, Japan, and Korea, as well as joint study of past issues and visits to such historic sites as the Nanjing Massacre Memorial Hall, Hiroshima Peace Memorial Park, and Seodaemun Prison History Museum in Seoul would prove valuable to developing comparative and ultimately more comprehensive views of the region's history. Such gatherings would be a regionwide attempt to share and heal pains of the past. By bringing history more to the forefront, they would, arguably, serve to reduce its relevance in national identity and bilateral relations.

Deng Xiaoping said, "It does not matter if this question is shelved for some time, say, ten years. Our generation is not wise enough to find common language on this question. Our next generation will certainly be wiser. They will certainly find a solution acceptable to all."[39] Contrary to his expectations, Northeast Asian nations find themselves in a worsening situation involving history and territorial disputes, and their history education tends to be more nationalistic. Today's youth, armed with universal values that underscore democracy, human rights, and peace, have a foundation capable of reducing the gap in national identities. Overcoming the history gap with Japan by placing the past in the context of universal rights, as championed by the United States, would be a force for refocusing each country's identity and promoting regional reconciliation. Three identity gaps placed in this context through an active US role can be addressed together.

Notes

1. "Transcript: President Park Geun-hye, Republic of Korea—Speech to Joint Session of Congress," May 8, 2013, http://woodall.house.gov/transcript -president-park-geun-hye-republic-korea-speech-joint-session-congress -may-8-2013.

2. Ankit Panda, "Park Geun-hye: Japan Summit 'Pointless' without Apology," *The Diplomat*, November 5, 2013, http://thediplomat.com/2013/11/park-geun -hye-japan-summit-pointless-without-apology/.

3. Yunbi Zhang, "Xi Warns Abe over Diaoyu Islands," *China Daily*, September 7, 2013, http://europe.chinadaily.com.cn/china/2013-09/07/content_16950986 .htm.

4. "Kishida Assures Kerry on South, Syria, WMDs," *The Japan Times*, October 4, 2013, http://www.japantimes.co.jp/news/2013/10/04/national/kishida-assures -kerry-on-south-syria-wmds/#.UsPrKPZQ3uU.

5. "Abe Attempts to Deny Japan's Invasion History," *CCTV*, May 9, 2013, http:// english.cntv.cn/program/newsupdate/20130509/101149.shtml.

6. "Seven of Ten Koreans Say Japan Has Not Apologized," *Asia Today*, November 21, 2013, http://www.asiatoday.co.kr/news/view.asp?seq=896846.

7. Dong-hwan Ahn, "94% Koreans Say Japan Feels No Regret for Its Past Wrongdoings, 63% Japanese Find Korean Demand for Japanese Apology Incomprehensible," *Seoul Shinmun*, January 4, 2013, http://www.seoul.co.kr /news/newsView.php?id=20130104001010.

8. See articles in Tsuyoshi Hasegawa and Kazuhiko Togo, eds., *East Asia's Haunted Present: Historical Memories and the Resurgence of Nationalism* (Westport, CT and London: Praeger Security International, 2008).

9. See Gi-Wook Shin and Daniel Sneider, eds., *History Textbooks and the Wars in Asia: Divided Memories* (London: Routledge, 2011).

10. Gi-Wook Shin, "History Textbooks, Divided Memories, and Reconciliation," in *History Textbooks and the Wars in Asia: Divided Memories*," in Gi-Wook Shin and Daniel C. Sneider, eds., *History Textbooks and the Wars in Asia: Divided Memories* (New York: Routledge, 2011), 14.

11. Ibid.

12. Richard McGregor and Simon Mundy, "Ill Will between Japan and South Korea a Strategic Problem for U.S.," *Financial Times*, November 21, 2013, http://www.ft.com/cms/s/0/d40e3b00-5232-11e3-8c42-00144feabdc0 .html#axzz2p8oPSZXk.

13. Martin Fackler and Sang-Hun Choe, "A Growing Chill between South Korea and Japan Creates Problems for the U.S.," *The New York Times*, November 23, 2013, http://www.nytimes.com/2013/11/24/world/asia/a-growing-chill -between-south-korea-and-japan-creates-problems-for-the-us.html.

14. Victor Cha, *Alignment Despite Antagonism: The United States-Korea-Japan Security Triangle* (Stanford, CA: Stanford University Press, 1999), 209.

15. Jong-hoon Ha, "Dokdo and History Questions as the Most Critical Problems of the Japan-Korea Relations," *Seoul Shinmun*, January 4, 2013, http://www .seoul.co.kr/news/newsView.php?id=20130104003004.

16. Alexis Dudden, *Troubled Apologies among Japan, Korea, and the United States* (New York: Columbia University Press, 2008), 5.

17. Wikipedia, "List of War Apology Statements Issued by Japan," http:// en.wikipedia.org/wiki/List_of_war_apology_statements_issued_by_Japan.

18. For more on Japanese conservatives' views of Asia, see Wakamiya Yoshibumi, *Sengo hoshu no Ajia kan* (Tokyo: Asahi shimbunsha, 1997).

19. Yoshiaki Yoshimi, "Government Must Admit 'Comfort Women' System Was Sexual Slavery," *The Asahi Shimbun*, September 20, 2013, http://ajw.asahi .com/article/forum/politics_and_economy/AJ201309200068.

20. For instance, see the newsletters by the Society for the Dissemination of Historical Fact in Japan. According to the SDHF, "An examination of those documents reveals that the provenance of accusations that Japan perpetrated a massacre in Nanking is wartime propaganda initiated by the Nationalist intelligence organization. They also expose European and American Nationalist agents who were intimately involved in the concoction of 'Nanking Massacre' propaganda." See: http://www.sdh-fact.com/CL02_1/27_S4.pdf.

21. Dudden, *Troubled Apologies among Japan, Korea, and the United States*, 33.

22. Daniel C. Sneider, "The War over Words: History Textbooks and International Relations in Northeast Asia," in Gi-Wook Shin and Daniel C. Sneider, eds., *History Textbooks and the Wars in Asia:* 246–268.

23. "North East Asia's Undercurrents of Conflict," report of the International Crisis Group, December 15, 2005, 13.

24. "Japan, S. Korea Researchers at Odds over Forced Labor, Comfort Women," *The Japan Times*, March 24, 2010, http://search.japantimes.co.jp/cgi-bin /nn20100324a3.html.

25. In his opening speech, Bu Ping proclaimed, "In Japan, speeches and activities not admitting the responsibility for the war of aggression and denying the historical

facts of the war have existed until now. Those irresponsible words and actions going against the common interests of the two countries have constantly hurt the public sentiment of a war victim nation." See reports in *Mainichi Shimbun* (December 27, 2006) and *The Financial Times* (February 16, 2007).

26. See *The Japan Times* (March 21, 2007) and *Xinhua* (March 21, 2007).

27. "China-Japan Scholars' Report Completed," *People's Daily*, February 1, 2010, http://english.people.com.cn/90001/90776/90883/6884274.html.

28. "New Study Fails to Bridge Japan, China History Divide," *AFP*, January 31, 2010, http://english.people.com.cn/90001/90776/90883/6884274.html.

29. Takashi Oshima, "U.S. Report Raises Concerns about Abe's Perception of Japan's Wartime History," *The Asahi Shimbun*, May 9, 2013, http://ajw.asahi.com/article/behind_news/politics/AJ201305090101.

30. David Straub, "The United States and Reconciliation in East Asia," in Tsuyoshi Hasegawa and Kazuhiko Togo, eds., *East Asia's Haunted Present* (Westport, CT: Praeger, 2008), 212.

31. Alexis Dudden, "Dangerous Islands: Japan, Korea, and the United States," *Japan Focus*, August 11, 2008, 2–3.

32. Gilbert Rozman, "Japan and Korea: Should the U.S. Be Worried about Their New Spat in 2001," *Pacific Review* 15, no. 1 (2002): 26.

33. Daniel C. Sneider, "A Dangerous Stalemate between Japan and South Korea," *The Washington Post*, October 31, 2013, http://www.washingtonpost.com/opinions/daniel-sneider-the-us-can-facilitate-healing-between-japan-and-south-korea/2013/10/31/d4db3d84-40b3-11e3-a751-f032898f2dbc_story.html.

34. McGregor and Mundy, "Ill Will between Japan and South Korea a Strategic Problem for U.S."

35. "Northeast Asia's Undercurrents of Conflict," International Crisis Group, December 2005, http://www.crisisgroup.org/home/index.cfm?id=3834.

36. Another example is the contrasting role of the United States in dealing with foreign forced labor: "The U.S. pressed hard to force the reluctant German government and corporations to admit their role, make a public apology to the aggrieved, and provide compensation. Toward the Japanese government, by contrast, the U.S. position was precisely opposite, protecting it against claims at every step, even before the San Francisco Treaty." David Palmer, "Korean Hibakusha, Japan's Supreme Court and the International Community: Can the U.S. and Japan Confront Forced Labor and Atomic Bombing?" *Japan Focus*, February 2008, http://japanfocus.org/products/details/2670.

37. Straub, "The United States and Reconciliation in East Asia," 215.

38. Gi-Wook Shin, "Perspective: Historical Disputes and Reconciliation in Northeast Asia: The US Role," *Pacific Affairs* 83, no. 4 (December 2010): 663–673. Offers suggestions on how the United States can play a constructive role in historical reconciliation.

39. Deng Xiaoping, then vice-premier of the People's Republic of China, at a press conference in Tokyo on October 25, 1978, answering a question about Diaoyu Island put forward by a Japanese journalist, quoted in "New Upsurge in Friendly Relations between China and Japan," *Peking Review* 41 (1978): 16.

CHAPTER 17

National Identity under Transformation: New Challenges to South Korea

Kim Jiyoon

Eleven years ago, observers of South Korea were stunned as they witnessed two critical events unfold in central Seoul. The first was one of reverie as the plaza of City Hall literally became a sea of red. People wearing red devil t-shirts, the mascot of the Korean national soccer team, gathered to root for the Korean national soccer team in the 2002 World Cup, cohosted by South Korea and Japan. The second was one of protest. In the same plaza, a huge mass again assembled in the winter to hold a candlelight vigil mourning the deaths of two junior high school girls killed by a US armored vehicle. The series of candlelight vigils generated public uproar against the perceived unfairness of the Status of Forces Agreement (SOFA) and sparked a wave of anti-Americanism. The incident created serious concerns about the ROK-US alliance and among the political elite of both countries. There was no shortage of scholarship depicting these two incidents as epitomizing the ethnic nationalism of Korea. They became prominent examples in a growing literature on Korean national identity.

Eleven years later, the situation has considerably changed. International relations surrounding the Korean Peninsula, the ideological position of the Korean public as a whole, and inter-Korean relations have all gone through a significant transformation. Most importantly, a new generation is coming to the fore. Although a strong sense of ethnic homogeneity is persistent among Koreans, the emergence of a new, immigrant demographic appears to be changing this. The total immigrant population in

South Korea is reported to surpass 3.1 percent of the total population in 2013 and it has been increasing for ten years.[1] Seeing a foreigner on the streets of Seoul is no longer an uncommon thing and the degree of exposure of South Koreans to new culture, new language, and new people is higher than ever. This internal societal shift is, I argue in this chapter, leading to a change in thinking about national identity, that is, "Koreanness."

Any transformation in the nature of Korean national identity is anticipated to bring about consequences relating to international relations. Ample research confirms that South Korean national identity is critical to understanding Korea's international relations, with countries such as the United States, China, Japan, and North Korea in particular. Most of these studies treat Korean national identity and the nationalism shared by the Korean people as being ethnically dominated. However, as the shape of ethnic and demographic composition is shifting, we find evidence of a far-reaching change in Korean national identity as well.

A change in national identity is expected to bring about a modification to the perspectives held on those four countries, and particularly North Korea—the country with which South Koreans share an ethnic bond. Furthermore, attitudes toward unification should differ from ten years ago, as the rationale compelling unification is fundamentally derived from staunch ethnic nationalism. This creates a new dilemma for Korea. It needs to decide between two policy choices—to maintain ethnic nationalism for the sake of the unification of the peninsula or to cultivate civic nationalism in preparation for a multiethnic Korea. Whatever leaders decide, the forces inducing change in national identity may not be responsive to their choices.

This chapter identifies indications of the transformation of national identity, how that is reflected in attitudes toward North Korea and on unification. I demonstrate the two types of nationalism that lead to a specific attitude toward a nation and its people. I examine changes in the perception of Koreanness using the Asan Institute's Daily Polls, noting a tremendous generational gap, as on other issues. I then delve into South Korean attitudes toward North Korea and unification, suggesting challenges in the foreseeable future.

The National Identity Framework Applied to South Koreans and Unification

National identity is a subject increasingly explored. One stream of scholars asserts that the nation is a created modern concept. Benedict

Anderson called a nation an "imagined community" that is socially constructed by those who believe in belonging to the group. It is practically impossible that all the people in a nation have the same ethnic origin and have maintained it for a long period. Another stream argues for a "*primordial*" type of nationalism and emphasizes that premodern ethnic roots are essential. Anthony Smith represents the "modernist" view by asserting that ethnic roots provide the pre-existing basis for a nation. Without it, he asserts, the durability and strong appeal of a nation can hardly be explained.[2]

Works on South Korean national identity are also classified into these contending frameworks.[3] Numerous historians and the public tend to believe that Koreans have been a single race from the prehistoric period. Believing the myth about *Dangun* and being descendants of him, the ethnic unity of Korea came naturally to many Koreans. As Shin notes, even the Three Kingdoms period has been described as continuous efforts to unify and restore one-nationness, which provides a strong mandate for the unification of the two Koreas.[4]

In contrast, constructivists such as Andre Schmid assert that the Korean nation is a product of nationalists in the late nineteenth century. Korean ethnic nationalism was created to separate Korea from China and to fit into the modern international system.[5] To these scholars, the fact that Korea maintained incredible territorial integrity for an extended period of time does not fulfill the condition for primordial ethnic nationalism. It was a rigid society with restrictive class strata. The Korean elites considered themselves as belonging to the civilized world of China rather than forming a nation with the commoners of the Chosun dynasty.[6] This implies the priority of a kind of Confucian civilizational identity, but with the Manchus ruling China and a growing sense of superiority in Confucianism, a case can be made that a sense of being separate from China was already strengthening a sense of Koreanness before this period.

Constructivists and primordialists agree that modern Korean nationalism emphasized ethnic unity and a shared bloodline. Shin accurately notes that "race, ethnicity, and nation were conflated" in Korean nationalism, seen in frequent usage of the term *minjok*, which sometimes implies ethnicity and at other times nation.[7] According to him, ethnicization of Korean nationalism was developed in the late 1920s in response to Japanese racial discrimination. After the ethnic nationalism was at its height, Korea was divided into two, which propelled competition for the legitimacy of the one ethnic Korea from Rhee Syngman and Kim Il-song. The late-1980s democratization movement helped to embed nationalism. As Katherine Moon points out, democratization was aided by nationalism epitomized by anti-Americanism.[8]

Omnipresent in Korean politics, South Korean nationalism is at the center of unification discourse as well. Because the country has always been perceived to be one ethnic nation, the current division of Korea has been regarded as temporary. Both South and North regimes claimed legitimacy over one ethnic Korea, accusing the other side of being a puppet regime of superpowers during the Cold War. Unification became a raison d'etre for both sides. Behind the Sunshine Policy of Kim Dae-jung and Roh Moo-hyun was Korean ethnic nationalism. Otherwise, no proper explanation is possible for providing economic aid to a country still technically at war with South Korea. Thus, the desire for unification has been largely driven by ethnic nationalism. This aspiration has had ramifications for other foreign policies.

Ethnic National Identity or Civic National Identity?

Even before the recent upsurge in foreigners living in Korea, observers were pointing to a rise in civic identity at the expense of ethnic identity.[9] It appeared to be a natural byproduct of modernization and globalization or of democratization, and it was being contested. Alford and Samuel Kim accurately portrayed how Koreans tailored the premise of globalization into a nationalistic agenda. According to Alford, at the arrival of the globalization theme, Koreans still strongly maintained ethnic identity and understood globalization on their own terms.[10] Similarly, Samuel Kim asserts that globalization or *segyehwa* in Korea was driven by the government, which did not result in any significant strides to make changes in the cultural nationalism of Korea.[11] While being supportive of the fact that no significant changes had been made, as Samuel Kim and Alford argue, Shin disagrees with the claim that there has been no interplay of globalization and nationalism. Specifically, Shin argues that nationalism "appropriated" globalization to increase national pride.[12] Nonetheless, the consensus by these authors is that globalization has not undermined ethnic nationalism, but enhanced it.

However much forces other than migration were playing a role, there is little doubt that the transition to civic national identity was proceeding more slowly than many expected. Ethnic national identity maintained vitality, even as Lee Myung-bak was showcasing "global Korea" and shared support for "universal values." While ethnic identity was being more seriously challenged and support for unification was no longer so widespread,[13] the impact of immigration was still gathering steam.

That is the focus of this chapter, which identifies one powerful force now reshaping national identity: increasing heterogeneity of the population.

It is becoming increasingly important to understand the impact the influx of immigrants will have, transforming long-held views of Koreanness. As of 2013, foreigners living in Korea exceeded 1.5 million—about 3.2 percent of the whole population. This is still small compared to the immigrant population of Western democracies such as Germany and France. However, the number is projected to continuously increase, reaching 10 percent of the population by 2030. Of the total, immigration by marriage constitutes about 20 percent. Many foreign women, mostly from China and Southeast Asia, move to South Korea to marry Korean men residing in rural areas. While international marriages make up approximately 10 percent of all marriages, more than 40 percent of marriages taking place in rural areas are international. This type of immigration creates the "multicultural family." The number of children from multicultural and multiethnic families has steadily increased to reach almost 200,000 in 2013.[14] It is even projected that 49 percent of all children in farming areas will be multiethnic by 2020.[15] Without a doubt, the long-held belief in ethnic homogeneity is on the verge of drastic change. Will it change the definition of Koreanness as perceived by the Korean public from ethnic oriented to nation or civic oriented? Some change has already occurred, but what would be the consequences of a sharp acceleration prompted by a much more heterogeneous population?

Anthony Smith asserts that national identity is formed by two components, "ethnic" and "civic."[16] Within a territory, all citizens enjoy the same rights and responsibilities under the common law. Civic culture is transmitted by education and the socialization process. The same ancestry, prehistoric myths, and memories play an indispensible role in forming the ethnic component of national identity. The civic and ethnic components are not mutually exclusive. In a nation, one component may overshadow the other, but on many occasions the two components coexist. In fact, Jones and Smith's analysis of surveys demonstrates that nationalism in many countries contains both components.[17]

Although it is believed that Korean national identity is mainly affected by the ethnic component, there is a possibility of modification due to the recent demographic shift. Kang Won-Taek and Lee Nae-Young's edited volume, *Understanding Korean Identity: Through the Lens of Opinion Surveys*, provides a helpful starting point for this discussion. Kang, in this book, notes that Korean nationalism has been maintained by an ethnic myth for a long period, but soon will be challenged by an ethnically transformed Korean society. He then asks what North Korea means to South Koreans in accordance with this change and expresses concerns about "South Korean nationalism" prevailing in the foreseeable future.[18]

I asked the same questions relating to "civic" and "ethnic" components in the 2013 Asan Daily Poll. The same seven questions on national identity used by Kang and Lee are asked for comparison. Of the seven criteria, I use three to measure the strength of a respondent's ethnic component: (1) being born in Korea, (2) having the Korean bloodline, and (3) living in Korea for most of one's life. The remaining four are used to measure the strength of one's civic component: (4) possessing Korean nationality, (5) being able to speak and write in Korean, (6) abiding by the Korean political and legal system, and (7) understanding Korean traditions. Table 17.1 presents the 2013 results compared with 2005 and 2010 from the Kang and Lee study. Following the Korean political and legal

Table 17.1 Preconditions for Koreanness

	Year	Important	Not important
Ethnic component			
Being born in Korea	2005	**81.9**	17.7
	2010	87.7	12.2
	2013	69.0	27.9
Having the Korean bloodline	2005	80.9	18.3
	2010	84.1	15.4
	2013	65.8	30.4
Living in Korea for most of one's life	2005	64.6	34.7
	2010	78.2	21.5
	2013	66.1	30.2
Civic component			
Maintaining Korean nationality	2005	88.2	11.1
	2010	89.4	10.5
	2013	88.4	9.10
Being able to speak and write in Korean	2005	87.0	12.6
	2010	87.8	12.2
	2013	91.7	6.70
Abiding by the Korean political and legal system	2005	77.5	20.6
	2010	87.3	12.4
	2013	**93.4**	4.20
Understanding Korean traditions	2005	80.9	18.3
	2010	85.9	14.0
	2013	91.5	6.10

Source: Asan Daily Poll (November 29–December 1, 2013).

system was identified as the most important condition to be recognized as Korean in 2013. Surprisingly, the least important factor was having the Korean bloodline, which is the sine qua non for ethnic nationalism.

Notable declines are detected in the number of those who agree with two statements: "a person should be born in Korea" and "having the Korean bloodline." In 2005 and 2010, being born in Korea was identified as important by 81.9 percent and 87.7 percent, respectively. However, there was a huge decline in 2013 with only 69 percent so indicating. In contrast, the number of those who think that nativity is *not* important more than doubled to 27.9 percent. A similar tendency is discovered in the statement that a Korean should have the Korean bloodline. The numbers were nearly 80 percent in both 2005 and 2010, dropping to 65.8 percent in 2013, when as many as 30.4 percent thought that sharing the same bloodline is not important to being Korean. The percentage who consider living in Korea for most of one's life is, importantly, relatively low (66.1 percent).

In contrast, the importance of elements related to civic nationalism grew or stayed at about the same level. Approximately 88 percent of respondents answered that keeping Korean nationality was important, which is only a 1 percent drop from the 2010 result. The ability to use Korean language remained important, with 91.7 percent of respondents in agreement. Two elements of civic identity are perceived to be more important than previously as a measure of Koreanness. In 2005, respecting the Korean political and legal system was regarded as essential for a Korean by 77.5 percent, and in 2013, by 93.4 percent, the highest response in the survey. Conversely, those who think that respect for the Korean political and legal system is not important for Koreanness dropped to 4.2 percent from 20.6 percent in 2005. The Korean public also increasingly perceives understanding Korean traditions as significant for being a Korean. While 80.9 percent of respondents considered it to be so in 2005, the number rose to 91.5 percent in 2013.

As seen in numerous cases, generational differences are found in the data (table 17.2). The gap is particularly discernible when it comes to the ethnic component of national identity. Apparently, the younger generations care much less about ethnic components such as nativity, bloodline, and residence. For example, only 55.5 percent of those in their twenties agreed that a Korean should share the same bloodline, while those in their sixties or over still consider the condition important at the rate of 81.5 percent. In addition, only 55.4 percent of respondents in their twenties think that being born in Korea is an important condition for being considered Korean, whereas as high as 82.4 percent of

Table 17.2 Preconditions for Koreanness: by age group

		Important	Not important
Ethnic component			
Being born in Korea	20s (186)	55.4	43.4
	30s (206)	63.3	35.1
	40s (221)	65.7	30.7
	50s (188)	78.4	19.1
	60s or over (199)	82.4	11.0
Having the Korean bloodline	20s (186)	55.5	42.4
	30s (206)	62.6	34.5
	40s (221)	58.2	38.3
	50s (188)	71.9	24.4
	60s or over (199)	81.5	11.7
Living in Korea for most of one's life	20s (186)	59.4	37.7
	30s (206)	59.6	37.0
	40s (221)	65.6	32.0
	50s (188)	72.0	24.6
	60s or over (199)	74.1	19.8
Civic component			
Maintaining Korean nationality	20s (186)	88.2	10.4
	30s (206)	85.7	13.3
	40s (221)	87.1	11.1
	50s (188)	90.8	6.90
	60s or over (199)	90.5	3.20
Being able to speak and write in Korean	20s (186)	91.3	8.10
	30s (206)	94.4	5.60
	40s (221)	89.1	9.30
	50s (188)	92.7	6.10
	60s or over (199)	91.0	4.20
Abiding by Korean political system and legal system	20s (186)	94.3	5.70
	30s (206)	92.9	5.40
	40s (221)	95.4	3.00
	50s (188)	93.8	4.30
	60s or over (199)	90.4	3.00
Understanding Korean traditions	20s (186)	88.1	8.90
	30s (206)	93.7	5.00
	40s (221)	86.7	10.6
	50s (188)	96.0	4.00
	60s or over (199)	93.4	1.50

Source: Asan Daily Poll (November 29–December 1, 2013).

the oldest generation thinks that way. A similar tendency is found in the question of living in Korea for most of one's life. Only 59.4 percent of Korean youth think that this is essential for Korean identity, versus 74.1 percent of the elderly who take this as essential.

Although there is a slight disparity across age groups, not as much difference is found in the civic component responses. The young generation generally agrees with the old generation's opinions on the civic components in Koreanness. What Korean youth thought most important to be Korean is following the rule of law in Korea and respecting the Korean political system. More of them (94.3 percent) than the elderly (90.4 percent) consider this important for being Korean.

Compared with the responses eight years ago, the percentage of Koreans who consider the ethnic components such as bloodline, nativity, and residence has decreased in 2013. While fewer people depend on ethnic elements to determine Koreanness, the importance of civic elements was maintained or slightly enhanced. This tendency is more visible among the young generations, who enjoy higher levels of exposure to foreign culture and are living in a more globalized environment. Although educated in the myth of one ethnic Korea and *Dangun* at school, the multiethnic community in which they are living requires a new concept of national identity and Koreanness. To them, civic components carry much greater weight than ethnic components for Korean identity. Gradually, but steadily, Korean nationalism is undergoing transformation, and it is starting from the younger generation.

The rise of civic nationalism is worthy of notice because in many other developed countries change is in the opposite direction as their society becomes multiethnic. With the influx of immigrants, numerous societies are becoming more ethnically nationalistic. The reason why South Korean youth are not following this path but tend to be more civically nationalistic can first be found in the small size of the immigrant pool. Until they become visibly threatening in the society in terms of size, the South Korean public is likely to cultivate civic nationalism. A second reason can be attributed to the considerable proportion of immigrants to Korea coming as a result of marriage. Marriages between bride migrants and South Korean guys residing in rural area generate multiethnic and multicultural families. The Korean government has made special efforts to embrace this rapidly growing type of family and children. In that regard, well-publicized propaganda distributed by the government to treat multiethnic and multicultural families well has facilitated accommodating civic nationalism instead of ethnic nationalism to a certain extent.

What Is North Korea to South Koreans?

In the previous section I found that the civic component of nationalism was gaining in significance, and it is more apparent among the Korean youth. If ethnic nationalism is gradually fading, particularly among the young generation, what would be the impact on the attitude toward North Korea and unification, whose principal driving force is ethnic nationalism?

Figure 17.1 indicates country favorability ratings measured on a 0–10 scale, with 0 indicating "zero favorability" and 10 indicating very favorable. The United States is ranked as the most favored country with a score of 6.4, while Japan was the least favored country with a score even lower than that of North Korea. The survey was conducted in September 2013. There were several North Korean provocations in the first half of the year, with the North's third nuclear test in February the most memorable. Thus, the public response to these incidents might have been reflected in the score. Still, North Korea was not a favorable country to many South Koreans.

What is more illuminating is the response to hypothetical soccer matches between countries. Gi-Wook Shin tested Koreans' favored country by posing the same hypothetical soccer matches. North Korea received disproportionately higher support from South Korean respondents in matches against the United States and Japan. For instance, in a hypothetical match between North Korea and the United States, 83.2 percent of respondents answered that they would root for North

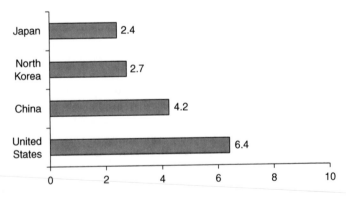

Figure 17.1 Country favorability ratings (2013).
Source: Asan Annual Survey 2013.

Korea and only 15.5 percent for the United States. In the case of Japan, 91.4 percent of South Koreans were more willing to root for North Korea than for Japan.[19]

A similar set of questions was asked by the Asan Daily Poll in October 2013. The result was quite different (table 17.3). In a hypothetical soccer match between North Korea and the United States, slightly more than a majority (51.2 percent) of South Koreans chose the United States over North Korea. In a game against Japan, North Korea continues to have more supporters. However, support for North Korea has decreased considerably from 91.4 percent to 77.7 percent.

When examined by age group, there is an interesting but consistent tendency. The young generation of South Korea exhibits conservative attitudes toward national security issues. They are quite a different species from the young generation ten years ago. Conventionally, a conservative South Korean tends to be hostile and assertive toward North Korea and friendly toward the United States. Much like those who are in their sixties, a disproportionate number of the youngest generation of Korea chose to support the United States (64.8 percent) in the hypothetical match against North Korea. This is the second highest proportion following the oldest generation's support 72.8 percent. The most ethnically bound generation was in their forties—the so-called 386 generation. "Cohort theory" applies handsomely.[20]

Table 17.3 Hypothetical soccer match

		North Korea	US
	Total	43.5	51.2
	20s	27.0	64.8
North Korea vs. US	30s	54.0	39.6
	40s	65.2	30.7
	50s	44.3	51.5
	60s or over	23.5	72.8
		North Korea	Japan
	Total	77.7	14.3
	20s	65.5	28.2
North Korea vs. Japan	30s	84.0	9.80
	40s	89.0	5.40
	50s	82.9	12.2
	60s or over	65.4	18.0

Source: Asan Daily Poll (October 6–8, 2013).

The young generation's differing mindset on nationalism is also found in their attitude toward North Korea and unification. Table 17.4 presents how Koreans view North Korea in 2013. The responses were classified into four groups: "as one of us," "as a neighbor," "as a stranger," and "as an enemy." A person under the influence of ethnic nationalism would be more likely to think of North Korea as one of us and look forward to unification as soon as possible. The age group most ethnically bound to North Korea is, again, the 386 generation, 36.4 percent of whom viewed North Korea as one of us. Without a doubt, the youngest generation has the weakest ethnic link to the North. Only 17.9 percent—half the proportion of that of the 386 generation—view North Korea as one of us. Also important is the high proportion (16.8 percent) who think of the North as a stranger, a term lacking an emotional element. Even enemy is more emotionally driven. The fact that quite a significant fraction of the Korean youth sees North Korea as a stranger indicates the emotional distance they feel from North Korea.

Opinions diverge when it comes to unification too (table 17.5). The young generation appears much less interested. Only 9 percent responded that unification should be achieved as soon as possible, the lowest number across all age groups. Almost the same percentage

Table 17.4 Perception of North Korea by age group

	One of us	Neighbor	Stranger	Enemy
20s	17.9	28.7	16.8	23.5
30s	25.5	34.9	12.1	16.8
40s	36.4	29.4	9.80	16.2
50s	33.6	28.8	3.80	26.0
60s and over	23.8	31.4	8.30	28.6

Source: Asan Annual Survey 2013.

Table 17.5 Attitude toward unification by age group

	As soon as possible	Depending on circumstances	No need to hurry	No need to unify
20s	9.00	60.1	21.6	9.30
30s	13.8	61.7	17.1	7.40
40s	21.1	61.5	13.8	3.70
50s	17.5	57.9	21.2	3.40
60s and over	21.0	51.1	23.5	4.40

Source: Asan Annual Survey 2013.

do not want to see unification (9.3 percent), a stark contrast to the 21.1 percent in their forties who wanted to see unification as soon as possible. Although those in their sixties or older were most hostile and assertive toward North Korea with nearly one-third of them perceiving it as an enemy, they tend to insist on the need to reunify as quickly as possible at a higher proportion (21 percent).

From the survey results, we can confirm the weakening ethnic ties with North Korea and decreasing call for immediate unification. Unification does not seem to be an aim for the youngest cohort, which obviously results from deteriorating ethnic national identity shared with North Korea. Attitudes toward North Korea and unification largely correspond to the gradually changing nature of Korean national identity.

Conclusion

National identities change, normally becoming less focused on ethnic homogeneity and more conscious of a shared civic community. This transformation has been observed in Korea for some time, but analysts were struck by the enduring hold of ethnic identity. Frameworks for the study of national identity point to various causes of change, but in the case of Korea they were slow to recognize the force of migration. After all, until recently, the number of Koreans born outside the country was much smaller than the numbers of immigrants in the population of other developed countries. As migration intensified, not only did it have a profound impact on how people think about Koreanness, it also reverberated in attitudes toward relations with the outside world, mediated through changing perceptions of the desirability of unification.

Current international relations in Northeast Asia exemplify how ethnic nationalism of Korea plays a significant role. National identity gaps, as Gilbert Rozman described them, have long been determining bilateral relations among South Korea, Japan, and China.[21] Nonetheless, the nature of national identity is not permanent. Through external shocks and demographic changes, Korean society has begun to experience a multiethnic environment, which is at odds with the one-ethnic-Korea principle. Adaptation of national identity is inevitable. The findings here suggest challenges ahead for South Korea on how to deal with foreseeable conflicts arising from exclusive ethnic nationalism against those who are from other countries. As seen in numerous advanced societies, social integration of different ethnicities poses many challenges. Without a significant break from traditional nationalism, by far one of the most important drivers for the unification of the two Koreas, how can the views of older ethnic Koreans be coordinated with those of the new Koreans? Is weakening ethnic nationalism a fortunate thing for Korea? As ethnic national identity loses its strength among South Koreans, the unification of South and North also loses its raison d'etre. As crucial as ethnic nationalism has been as the foundation in relations with North Korea and unification policy, the rise of civic nationalism will inevitably dilute efforts for that. Is rethinking unification policy coming to the fore?

Notes

1. Korea Immigration Service *Monthly Statistics*, November 2013, http://www .immigration.go.kr/HP/COM/bbs_003/ListShowData.do?strNbodCd=noti 0097&strWrtNo=128&strAnsNo=A&strOrgGbnCd=104000&strRtnURL= IMM_6070&strAllOrgYn=N&strThisPage=1&strFilePath=imm/.
2. Benedict Anderson, *Imagined Communities* (London: Verson Editions and New Left Books, 1991); Ernest Gellner, *Nations and Nationalism* (Oxford: Blackwell, 1983). Clifford Geertz, "The Integrative Revolution: Primordial Sentiments and Civic Politics in the New States," in Clifford Geertz, ed., *Old Societies and New States: The Quest for Modernity in Asia and Africa* (New York: Free Press, 1963); Anthony D. Smith, *National Identity* (Reno: University of Nevada Press, 1991).
3. Gi-wook Shin succinctly summarized discourse on South Korean nationalism in his book *Ethnic Nationalism in Korea: Genealogy, Politics, and Legacy* (Stanford, CA: Stanford University Press, 2006).
4. Shin, *Ethnic Nationalism in Korea*.
5. Andre Schmid, *Korea between Empires, 1895–1919* (New York: Columbia University Press, 2002).
6. Shin, *Ethnic Nationalism in Korea*, 5–6.
7. Ibid., 4.
8. Katharine H. S. Moon, "Korean Nationalism, Anti-Americanism, and Democratic Consolidation," in Samuel S. Kim, ed., *Korea's Democratization* (New York: Columbia University Press, 2003).
9. Chung-in Moon, "Unravelling National Identity in South Korea: *Minjok* and *Gukmin*," in Gilbert Rozman, ed., *East Asian National Identities: Common Roots and Chinese Exceptionalism* (Washington, DC and Stanford, CA: Woodrow Wilson Center Press and Stanford University Press, 2012), 219–237.
10. C. Fred Alford, *Think No Evil: Korean Values in the Age of Globalization* (Ithaca, NY: Cornell University Press, 1999).
11. Samuel S. Kim, *Korea's Globalization* (Cambridge: Cambridge University Press, 2000).
12. Shin, *Ethnic Nationalism in Korea*.
13. Gilbert Rozman and Andrew Kim, "Korean National Identity: Evolutionary Stages and Diplomatic Challenges," in Gilbert Rozman, ed., *East Asian National Identities* (Stanford, CA: Stanford University Press, 2012), 197–217.

14. Ministry of Security and Public Administration, 2013, http://www.bokjiro
.go.kr/cmm/fms/FileDown.do?atchFileId=6115215&fileSn=67.5%20KB.

15. Won-Taek Kang and Nae-Young Lee, *Hankook-in, Urinun Nooguinka? Yeoron Josarul Tonghae Bon Hankook-inui Jeongcheseong* (Seoul: East Asia Institute, 2011).

16. Smith, *National Identity*; Anthony D. Smith, "The Myth of the 'Modern Nation' and the Myths of Nations," *Ethnic and Racial Studies* 11, no. 1 (1988): 1–26.

17. F. L. Jones and Philip Smith, "Individual and Social Bases of National Identity: A Comparative Multi-Level Analysis," *European Sociological Review* 17, no. 2 (2001): 103–118.

18. Kang and Lee, *Hankook-in, Urinun Nooguinka? Yeoron Josarul Tonghae Bon Hankook-inui Jeongcheseong.*

19. Gi-Wook Shin, "South Korean Anti-Americanism: A Comparative Perspective," *Asian Survey* 36, no. 8 (1996): 787–801.

20. Karl Mannheim, *Essays on the Sociology of Knowledge* (London: Routledge and Kegan Paul, 1952).

21. Gilbert Rozman, ed., *East Asian National Identities.*

Sources

Jang, Won Joon. 2009. "Multicultural Korea." *Stanford Journal of East Asian Affairs* 10, no. 2 (2009): 94–104.

Lee, Jee Sun E., "Post-Unification Korean National Identity," *Working Paper Series* 09–03 (2009), US-Korea Institute at SAIS.

Rozman, Gilbert. 2010. "South Korea' National Identity Sensitivity: Evolution, Manifestations, Prospects," *Academic Paper Series on Korea* 3 (2010). Korea Economic Institute.

CHAPTER 18

A National Identity Approach to Japan's Late 2013 Foreign Policy Thinking

Gilbert Rozman

Japanese national identity drew more attention in 2013 than at any time since the collapse of the bubble economy accompanied by the discrediting of *Nihonjinron* (the theory of the inherent superiority of Japanese culture) in the early 1990s. For the first time since 1945 Japan had a prime minister, Abe Shinzo, both unabashed in his commitment to transform national identity and optimistic that recent elections and public opinion polls give him a mandate to achieve this goal. More than at any time since they had normalized relations with Japan in 1965 and 1972, respectively, South Korea and China were obsessed with the negative nature of Japanese national identity. While taking satisfaction in what is viewed as Japan's unprecedented embrace of realism, loosening the grip of the pacifist streak in national identity more than six decades old, US officials were unavoidably drawn into the disputes over Abe's rhetoric and symbolic gestures, nervous about their impact on ties to neighbors beyond anything seen in recent East Asian history. At a time when advocates of realist theory might have been claiming vindication that rising tensions over territorial disputes and arms buildups confirmed their predictions and liberal theorists might have been salivating about accelerating tendencies to establish the TPP and a China-South Korea FTA, constructivist theorists pointing to the potency of elements of national identity were taking center stage in the analysis of international relations in East Asia linked to Japan.

This chapter applies a national identity framework comprised of six dimensions to debates in Japan at the end of 2013 most relevant to foreign policy. It differentiates four schools of thought in Japan seeking to reshape national identity. As case studies in how a struggle over national identity is playing out, three themes are highlighted. The conclusion puts the focus on a pending showdown between two directions in reshaping Japan's identity with important implications for intensifying competition between what is seen as a US-led Asia-Pacific community and a China-led East Asian community. Japanese pursue a great power national identity with divisive impact even as polarization in Asia deepens.

The six-dimensional framework in this chapter was developed in a series of three books.[1] The dimensions covered are: ideological, temporal, sectoral, vertical, horizontal, and intensity. The schools of Japanese thought differentiated for this analysis are: statism (*kokkashugi*), ethnic nationalism (*minzokushugi*), internationalism (*kokusaishugi*), and pacifism (*heiwashugi*). The case studies selected from recent Japanese debates are: (1) national security; (2) twenty-first-century economic order; and (3) the role of values in diplomacy. Coverage centers on writings in the final month of 2013, a time of unprecedented post-1945 efforts to prioritize national security, of intense negotiations to reach a 12-country agreement on TPP, of assertive Chinese moves such as establishment of an Air Defense Identification Zone (ADIZ) overlapping with territorial claims of Japan and South Korea, of deepening US frustration over the irreversible impasse in ROK-Japan relations, and of Abe's defiant year-end visit to the Yasukuni Shrine that aroused vehement denunciations from China and South Korea and, of no less salience, an expression of US disapproval. In the beginning of 2014 the fallout from Japan's struggle over identity was unrelenting.

The National Identity Framework

National leaders, media, and public opinion express views about what makes their state distinctive and superior. These views matter for domestic politics and also for relations with other states, which are deemed relevant as contrasts or impediments in realizing the symbolic manifestations of one's national identity. For Japanese, apart from the United States, China and South Korea have the most significance for national identity. In turn, these countries make Japan a focus of their national identities. This puts the spotlight on symbols of identity such as: the Yasukuni Shrine, compensation and rhetoric on "comfort women," territorial disputes linked to Japan's era of expansionism, and textbook wording on historically sensitive topics related to this era. As tensions have mounted with China and South Korea, these symbols of identity have been acquiring increased significance. Even if there are other causes of tension, the identity symbols capture the spotlight.

The struggle over Japan's national identity extends well beyond these specific symbols. It proceeds under the shadow of a battle over ideology—one dimension of identity—dating back to the 1950s, when Prime Minister Kishi Nobusuke, the grandfather of the current prime minister, pressed to overturn the ideology associated with pacifism and rejection of the course of history over more than half a century to 1945 with an ideology of restored veneration of Japan's aspirations in this period, including its goal of liberating Asia from Western imperialism. The postwar era saw heiwashugi continue as the dominant mode of thinking, and it still exercises considerable influence, even as it has lost ground in stages since the end of the Cold War. In contrast, the ideology found in kokkashugi unfettered by divisive civil society or international pressure was kept alive by a coterie of ardent conservatives and has been regaining influence. The struggle between these two clashing ideologies led to a stalemate, which some interpreted as a sign that Japan was just driven by pragmatism and avoided ideological themes, even "nationalism" in general. Behind the facade of a nonideological Japan were two forces hostile to realism and liberalism. What many saw as pragmatism in a country supposedly lacking a sense of its identity, in retrospect, was a standoff between two extremist ideologies bound to arouse problems.

The other five dimensions of national identity were discussed at length in the book *East Asian National Identities*, but can be highlighted for their impact on thinking that affects the struggle in 2013 between four schools of thought on three issues of contention. On the temporal dimension there is a fundamental divide over how to interpret the era to 1945 and the postwar era too. Should pride in the former era be renewed or wariness toward that era reaffirmed? As for the latter era, is it best seen as a tarnished interlude when full sovereignty could not be exercised or as an exemplary period when Japan achieved a lot and forged a foundation that can be sustained? Not only in bilateral relations, but also in internal divisions over identity, historical memory looms in the forefront of debates. In evaluating the post–Cold War period, there is a similar divide. Is Japan suffering from too little national pride or is it failing truly to embrace the US-led international community?

On the sectoral dimension, Japan has found little cause to reaffirm economic identity as a model and has upgraded political national identity in contrast to China without finding a way to differentiate itself from the United States as a champion of universal values, but cultural national identity survived the collapse of the bubble economy with potential for reinvigoration. Whereas at the peak of Nihonjinron boosterism, much was written about the unique qualities of individual Japanese and groupism, more emphasis is now given to the cultural characteristics of the state and the national community. One school holds that these have been neglected and need to be fostered by moral education and media support, as in a more assertive public persuasion role for the public NHK television station. At the opposite end of the spectrum is fear that claims of cultural uniqueness will be abused, as in the 1930s–1940s, by a leadership that does not respect the essence of postwar democracy. Cultural claims can be limited, not as extremist as Nihonjinron was, or they can be tied to state-centered nostalgia over the pre-1945 era, going much further in reclaiming the past.

One of the most intense battlegrounds over identity is the vertical dimension. This refers to how a country's internal organization is prized as distinctive. For some, a strong state led by politicians, not fragmented bureaucrats, is the essence of the desired national identity. They share with realists the goal of strengthening state capacity to meet the challenges of national security, but they proceed further in minimizing civil society for its right to information and to check state power. For others, the revival of a leviathan is equated with wartime authoritarianism and an identity at odds with the strong limits put on the state that became the foundation of Japanese national identity in postwar decades. The two sides stand at opposite extremes on the powers to be entrusted to the state.

The horizontal dimension weighs closeness to the United States, reentry into some sort of Asian community, and identification with an international community associated with the United Nations. Those who favor a closer US alliance do not necessarily seek to identify more with a country long seen as stifling Japan's own national identity, while those with hopes for more Asian regionalism struggle with new concern about dependency on China and lingering uncertainty about how Japan-ROK relations fit into the picture. References to Japan as part of the West have declined since the 1980s. Despite a last gasp of talk of an East Asian community under Hatoyama Yukio in 2009, that notion has also faded. If clarity on this dimension is lacking, its importance for identity remains undiminished, particularly with the United States and China battling for hegemony. Increasingly, the dividing line is between internationalism, accepting the United States more on identity terms, and kokkashugi, rejecting East Asia and, without fanfare, a US-led community too. This dimension is about Japan's place in the outside world: an integrating role with the United States, a divisive regional community with China, or a separatist role by itself.

The intensity dimension is divisive, as some Japanese, led by Abe, seek a much greater level of intense pride in their country, whereas others fear anything of the sort. Schools of thought vary on this dimension, as on the others noted earlier, and approaches to the hot topics of our time are heavily influenced by different attitudes toward identity intensity. At its core, this is a divide between those who would prioritize pragmatic diplomacy and an informed public for a "responsible" state, versus those who seek to arouse alarm that there is no trustworthy partner for Japan, whose citizens must be mobilized by identity. Staunch conservatives have been pressing hardest for more intensity in national identity.

Four Schools of Thought

On the right fringe of the Japanese political spectrum, there is support for kokkashugi, prioritizing the vertical dimension of identity under the sway of a lingering ideology. It is the US occupation that undermined this feature of Japanese society, weakening the state in the 1947 Constitution and through reforms aimed at decentralization. Thus, advocates of this school favor reduced dependency on the United States, constitutional reform as an end in itself, glorification of the state and emperor, and even a nuclear Japan. They regard bureaucracy as standing in the way of a top-down system, insist on moral education that puts loyalty to the state and state interests first, and rage against newspapers, officials, or even academics who are viewed as working against state authority. Criticisms of Abe are sometimes seen through this lens, as in an article in *Sankei Shimbun* warning that doubts in the United States about Japan becoming "*kokkashugiteki*" must be set aside, as Japan seeks its own path to defend its national interests.[2] While generally approving of Abe's leadership, advocates of this viewpoint find fault with his overreliance on the United States and put pressure on him to be less cautious. On the horizontal dimension, they warn that despite talk of "rebalance" the United States does not prioritize Asia, Obama has shown weakness on Iran and North Korea, and during his visit Joe Biden was less than forceful in supporting Japan against China's ADIZ, following the decision to allow US civilian aircraft to give advance notice to Chinese authorities. One *Sankei Shimbun* article even raised the Munich analogy in its criticisms of US appeasement.[3] Reluctant to criticize Abe, this school sought to prod him to be less cautious. It greeted his Yasukuni Shrine visit on December 26 with exhilaration, confirming convictions that Abe is one of them.

While a clear line between kokkashugi and minzokushugi is hard to draw and advocates of the former are also supportive of the latter, minzokushugi is less insistent on the state as separate from the society and more accepting of a realist US alliance. It prioritizes a distinct cultural identity centered, to be sure, on a cohesive state, but one rooted more on a homogeneous society than on state dominance over the society. Thus, as long as history is reevaluated to lift pride in the Japanese people at one with the state, there is room for affirmation of universal values and reinforcement of the US alliance without insistence on Japan's autonomous military power. The sectoral and temporal dimensions are most important, not the vertical and horizontal dimensions, and ideology plays a lesser role. At the same time, this is also an intensely conservative identity, for example, not amenable to realist compromise with South Korea over historical memory. Articles in *Sankei Shimbun* and in *Bungei shunju* at times reflect this thinking, but it is best found in *Yomiuri Shimbun* with its wider appeal, keeping more in mind the big picture and the valued US alliance.[4]

Kokusaishugi has shed much of the idealism associated with progressive thinking in the Cold War era to combine internationalism with realism for our times. It has lost a linkage to pacifist ideology in favor of a balance of power perspective and endorsement of both universal values and responsibility to help resolve international crises and threats. This means that the horizontal dimension is in the forefront with more room on the vertical dimension for civil society, including an active environmental movement and support for better opportunities for women. A more open Japan has room for foreign firms, leaving Japan to pursue its comparative advantage in certain industrial sectors. Accepting close ties to the United States, Japan's internationalists would broaden the range of global and regional activism in support of a US-led international community. This perspective can be found, to some extent, in all of the major Japanese newspapers, but rarely in *Sankei Shimbun*. *Asahi Shimbun* on the progressive side has shifted some in this direction. Yet, if we look for sustained "realist internationalism," none of the major newspapers endorses it unreservedly without qualifications linked to other national identity concerns. Perhaps, the closest to this thinking, centered on economic issues, not realism, is *Nikkei Shimbun*.

Advocates of heiwashugi seemingly have been marginalized, but they remain a major force in the media and had enough clout to help shape the policies of the DPJ in the first year it formed the government. They are wedded to the existing Constitution, have been energized by the movement against nuclear energy since 3/11, and insist on the postwar order as protection against unchecked state authority and military assertiveness. At its core, this is an ideological movement with strong views about the vertical as well as the horizontal dimensions. Its temporal identity stands in sharp contrast to that of the first two groups. *Tokyo Shimbun* may have taken the lead from *Asahi Shimbun* in this group with *Mainichi Shimbun* still showing some legacy of its past association here. Yet, it is hard to find the isolationist, Asian idealist forms of pacifism once prevalent in these newspapers, as all have made some accommodation with realist internationalism and should not be presumed to show no sympathy to minzokushugi. It is kokkashugi that looms as their chief nemesis, just as the reverse holds for the boosters of kokkashugi.

The two extreme schools tend to view the national identity conflict in dichotomous terms. They exaggerate each other's presence, making it seem as if their approach is the only alternative. Japan lacks journals as well as newspapers that cover international relations in a straightforward manner, putting internationalism at the center. No strong academic centers or think tanks have developed without a narrowly economic focus. Stifling civil society, bureaucracy has laid the groundwork for a skewed interpretation of the vertical dimension amenable to kokkashugi, while heiwashugi skewed the horizontal dimension. In failing to reconcile with South Korea, Japan has shown how identity harms diplomacy.

The Topics of Debate

National Security

National security used to be an afterthought in Japanese media coverage, but it is now the principal concern and the prime battleground for contending views of national identity. It is a matter of ideological significance for both extremes of the spectrum, and it puts the horizontal dimension in the forefront. Debate swirled around the decision to establish a National Security Council (NSC), a secrecy law that tightens control over information, and a new constitutional interpretation of collective defense. In the foreground are views of a possible "China threat" and of the dependability and desirability of the US alliance, as it is being strengthened. Given the centrality of pacifism and Article 9 as a barrier to full-fledged reestablishment of a "normal" military, we should not be surprised that views on national security test not only realist thinking but also Japan's broad national identity.

The kokkashugi camp supported maximal changes in Japan's security posture, expressing ambivalence about reliance on the United States. *Sapio* asserted that even if the US Department of Defense is seen as vigilant, the State Department is not, and the Biden early December visits to the region left Japan isolated, given US weakness in dealing with China.[5] *Sankei Shimbun* bemoaned: Susan Rice's November 20 speech, in which she used the terminology of a "new type of great power relations," a US inclination to avoid standing with Japan and to treat China equally, and its weakness in interpreting the Security Treaty with Japan.[6] Emphasizing Iran and the Middle East and seeking China's assistance in keeping North Korea in line, the Obama administration, lacking high-level officials versed in Japanese affairs, is of questionable reliability. In its final years the Bush administration had also been viewed with suspicion for its China and North Korea policies. For kokkashugi, however, what matters more is for Japan to defend itself, in the process strengthening its state and the identity centered on the state lost in 1945. As Abe's desire to rewrite the Constitution was frustrated, passing a secrecy law that tilted the balance sharply to the state and away from the media and the academic community served a palliative role.

The minzokushugi camp had a more limited focus on collective defense and historical revisionism, which it could reconcile with the embrace of universal values and the US alliance. Giving more weight to realist thought, it recognized that Japan could greatly increase its defensive capabilities through widening alliance or quasi-alliance relations with India, Australia, and other states in coordination with the United States. It clearly acknowledged that Japan cannot defend the existence of the nation by separate defense alone, welcoming reinterpretation of the Constitution to support the international order through rearguard assistance, arms exports, and other forms of collective defense.[7] Such proposals reflect the rapidly worsening security environment for Japan, resulting even in fear of attack missing in the Cold War. Yet, calls to place Japan within "international society" came with insistence on greater patriotism, boosting the NSC established in December but also ethnic pride,[8] ruling out a conciliatory posture toward South Korea.

Realist internationalists were most in line with US thinking about widening the scope of defense partnerships and alliances while continuing to try to engage China. Cognizant of Japan's reduced status, although few were willing to demote it from a great to a middle power, they put horizontal identity first without letting other dimensions, such as sectoral national identity, interfere. They firmly supported the new NSC, the secrecy law, and the alliance, as did the backers of minzokushugi. Yet, they weighed universal values as well as international responsibility more highly, and some expressed misgivings about the way the secrecy law was drawn to thwart an informed citizenry. Representative of this view is Kitaoka Shinichi, who calls on Japan to be a global player and assume a greater role in maintaining international and regional order. Acknowledging that it failed to play this role when it was only an economic great power and, at times, leaked military technology, he traces changes to the Persian Gulf War and recognition that Japan could contribute more to collective defense through the United Nations. Regretting how the United States mistakenly made the war on terror a crusade to make Iraq democratic and the resultant loss of support for Japan, which had stuck closely to it as well as had its image suffer as a democratic state through Koizumi's visits to the Yasukuni Shrine, Kitaoka keeps the focus on how best to counter rising security threats. Suggesting that the US priority for North Korea is nonproliferation of WMD rather than Japan's priority on nuclear weapons and missiles (not abductions), Kitaoka frets that the United States is concentrating on Iran and not North Korea. Warning that Japan cannot rely solely on the US nuclear umbrella, he still argues for boosting US ties, broadening collective defense not just with the United States, and boosting international society in the face of China's drive for a hegemonic, sinocentric, regional, and world order.[9] It is assumed that Yachi Shotaro, the first head of the NSC and key foreign policy advisor, has been pulling Abe toward kokusaishugi.

The heiwashugi camp was reinvigorated with the attack on the secrecy law led by *Asahi Shimbun*. They see security being used as a wedge to change Japan's vertical identity, as patriotism is prioritized in the name of peace and security.[10] They warn that Abe is using Japan's ally to change Japan's national identity in an ideological direction focused on the temporal and vertical dimensions. When the popularity of the Abe cabinet slipped 10 percent in a month as the secrecy law and fears of revisionism began to undercut earlier support that was generated through Abenomics, *Asahi Shimbun* saw problems likely to pile up in the spring of 2014 as the public sees the impact of a rise in the consumer tax and also a shift to collective defense.[11] Above all, as when Abe's popularity fell before he resigned as prime minister in 2007, the public would be fearful of his intentions to strengthen the state to the point of an arbitrary government. In an apparent reversal regarding the US role, advocates of this posture now tend to see it as a constructive force in trying to bring Japan and South Korea closer as well as to calm Sino-Japanese tensions. Even so, views of the secrecy law, collective defense, and the Futenma base issue all cast doubt on the wisdom of doing what those in other camps regard as essential for a closer US alliance.

Twenty-First-Century Economic Order

Negotiations over Japan's entry into the TPP have rekindled debate over how distinct is Japan's economic identity. If TPP is seen as a pathway to an economic order that forces China to change economically and in other respects, then the old dichotomy of a distinct Japanese economy with a nexus to its unique society and a US-driven global economic order threatening to Japan no longer is in the forefront. Indeed, with its expected impact on state-owned enterprises, services, subsidies, and investments, TPP is deemed to have considerable potential to transform Japan as well as China. This elicits an identity debate.

For the kokkashugi school, the US pressure on Japan's sensitive agricultural sector is excessive, related to Obama's desire to get results prior to the 2014 elections.[12] In not backing away from a 100 percent deal, as opposed to Japan's support for a 93 percent one with five agricultural products protected, Washington supposedly has surprised Tokyo with the expectation that Abe has wide popular support and can break Japan's taboo, as opposed to the view that Japan is needed for a deal, and Washington will have to yield.[13] After all, in the KORUS FTA it allowed for a 20-year delay in opening the South Korea market to pears and apples. Yet, even a *Sankei Shimbun* editorial acknowledged that Abe's level of support would fall if there is no TPP agreement and that his initiative to Southeast Asia and call for a new Asian rules-based structure would be damaged.[14] Others warn of the danger of TPP, bringing genetically modified crops, polluting Japan, and undermining national security in food.[15] The minzokushugi school is also wary about an opening of Japan that threatens the cultural nexus of farming and rural society. Given Japan's long economic decline and China's rise, however, an international identity has new appeal.

For *Nikkei Shimbun* and even much of the coverage in *Yomiuri Shimbun*, an axis of free trade would revitalize the Japanese economy. After years of stagnation, this is a way to bring a market opening to Japan.[16] Indeed, to achieve internationalism, joining this pact led by the United States and standing in contrast to China's increasing economic pressure is regarded as a high priority. Now that South Korea has expressed a desire to join TPP, fearing that Japanese liberalization would give it a competitive edge, there is hope of spurring exports to that country, where domestically produced cars still are protected while Japan is already quite open to its industrial exports. A TPP deal could accelerate China-Japan-Korea FTA talks, both because Japan and South Korea would join a high-quality FTA, but also because China may recognize that it too should join, as seen in its shift in attitudes from May 2013.[17] Yet, *Yomiuri Shimbun* puts the onus on the United States to show flexibility, warning of a weak Obama under pressure from Congress.[18]

Asahi Shimbun faults the US strategy for TPP and says that it was the loser by not getting a deal in 2013, but that Abe was also a loser because TPP is linked to Abenomics, which is the pillar of his support. In calling for a new strategy, it seems satisfied with a lower level agreement.[19] Heiwashugi long claimed the mantle of internationalism, but it was an idealistic version distrustful of the United States, not a welcome embrace of liberalism.

The Role of Values in Diplomacy

Aso Taro in 2006, foreign minister to Abe, called for an "arc of freedom and prosperity," and, although the term is not used today, Abe is pressing for a values-based diplomacy. Although this is viewed as nonproblematic for Japan-US ties and has been applied most vigorously to diplomacy in Southeast Asia with India another target, attention centers on how such advocacy relates to thinking on China. Values are a double-edged sword: under assault for its revisionist values, Japan's aggressive advocacy of universal values rings hollow, while appearing often to be unnecessarily confrontational toward China. One finds a division in Japan in how to treat values as a basis of foreign relations, with Abe stressing the rejuvenation of moral education, a distinct type of capitalism, and a sharp turn away from what he calls the DPJ's three years of "diplomacy of defeat" with values diplomacy given a big role in Japan's revival,[20] as if this is support for universal values.

The kokkashugi camp emphasizes values centered on the state while drawing a sharp line with values advocated in other countries. After years when the "Korea wave" featured in TV dramas gained wide following, leading many to think that Japanese and South Korean societies share much in common, the tide has turned with books on "hating the Korea wave" and even a December 2013 book focused on demonizing

South Korean society as not deserving its past positive image and fostering a civilization alien to that of Japan.[21] The main target, however, is postwar Japan supposedly imbued with hatred of the state. The Nihonjinron books glorified Japan's cultural traditions; the new literature is mostly about praising its state traditions. Terms such as "*bushido*" (samurai values) and "*seishin*" (spirit) are frequent in recently popular books on the virtues of Japan's "*kokka*" traditions.

Abe's followers consider unrestricted "worshipping" at the Yasukuni Shrine as a symbol of the end of the abnormal postwar era and prize the attached *Yushukan* museum for its historical narrative at odds with postwar temporal identity. Amending the Constitution is seen as necessary since its purpose had reputedly been to break the spirit of the Japanese nation. Abe in the spring of 2013 alarmed observers who feared fierce identity clashes. There was even alarm that the battle over historical memory would spread across the United States, driven by Korean Americans and Chinese Americans in a public relations battle that Japan was losing. In the backlash to his spring assertiveness, Abe grew quiet about historical memory as he met with more approval when he stuck to universal values. His visit to Yasukuni, however, pleased the kokkashugi and minzokushugi camps.

The left has long been suspicious of the values of internationalism as advocated by the United States, searching for a different path such as antinuclear ideals for a leadership role. Advocacy of an East Asian community was an alternative way to idealistic goals, but it too has lost traction. No ready choice for diplomacy is still open. Clinging to the Constitution as the source of values opposed to Abe's is the course remaining, although advocating new values, such as environmentalism, for our times, draws some support. At year-end polarization reigned: Abe's visit to Yasukuni mobilized the left, who conflated realist internationalism with statism, and Abe's base pushed for a more intense identity.

Conclusion

Abe Shinzo's electoral triumphs and popularity along with deepening concern over the "China threat" have tilted the debate over foreign policy steeped in national identity. The conservative camps of kokkashugi and minzokushugi have both been emboldened. They see a long-sought opportunity to bury heiwashugi and ride the tide of internationalism, limiting its realist cast. Serving as the driving force in identity debates since the LDP's return to power in the mid-1990s, they have reinvigorated identity after a loss of intensity in the first half of the 1990s.

A new peak in identity intensity in 2013 brings a possibility for ideology to escape from the shadows, for temporal identity to make a sharp break with the postwar era, for sectoral identity to make a comeback after its retreat from the spike in the 1980s, for vertical identity to draw on the pre-1945 legacy, and for Japan to reposition closer to the United States and in defiance of China with growing pretensions about Japan's leadership role in a values-laden global community. Conservatives in an uneasy alliance with internationalism on security and the economy are now emboldened.

On national security, realist internationalism has the upper hand, but doubts about the US "rebalancing" and determination to stand firm against China and North Korea leave room for the conservative schools to make their case. On TPP, internationalism has found new impetus, even if protectionist forces can resonate with their identity arguments to keep the pressure on the United States to relax its high standards. Even on values diplomacy, more fulsome embrace of universal values bodes well for the forces of internationalism despite renewed insistence on revisionist symbols. While antinuclear energy and antisecrecy law movements keep heiwashugi views in sight, the primary struggle is being fought within the Abe administration over putting kokkashugi first, not kokusaishugi.

The right-center coalition in Japan obscures a right versus center divide over national identity and policy choices. If a compromise is reached over TPP with some protectionism that can be construed as a favor to minzokushugi forces and the NSC finds a way to articulate Japan's distinctive foreign policy even as it cozies up further to the United States, forces on the right will take some solace, despite having largely to defer to the internationalists. As long as they are not constrained in warning about the "China threat" and North Korea, they can concentrate their case on the vertical dimension or how Japan must change in response to a more dangerous environment. At the crossroads of the two forces tugging national identity in somewhat different directions is thinking about South Korea. To see it as a threat to Japan's identity leads in one direction, while to seek to remove any identity considerations except shared universal values from this relationship would give a boost to realist internationalism. In the background is the degree to which Japan recognizes the US alliance as more than an expedient response to danger, deserving more convergence in values and a narrowing of the national identity gap with ramifications also for Japan's policy toward South Korea. Abe's obsession with identity is bringing an early showdown.

Notes

1. Gilbert Rozman, ed., *East Asian National Identities: Common Roots and Chinese Exceptionalism*; Gilbert Rozman, ed., *National Identities and Bilateral Relations: Widening Gaps in East Asia and Chinese Demonization of the United States*; and Gilbert Rozman, *The Sino-Russian Challenge to the World Order: National Identities, Bilateral Relations, and East vs. West in the 2010s.* All are published in Washington, DC and Stanford, CA by the Woodrow Wilson International Press and Stanford University Press in 2012, 2013, and 2014, respectively.

2. *Sankei Shimbun*, December 2, 2013, 7.

3. *Sankei Shimbun*, November 28, 2013, 7.

4. *Yomiuri Shimbun*, December 4, 2013, 3.

5. *Sapio*, January 2014, 1.

6. *Sankei Shimbun*, December 4, 2013, 4; *Sankei Shimbun*, December 8, 2013, 1.

7. *Yomiuri Shimbun*, December 18, 2013, 1.

8. *Yomiuri Shimbun*, December 18, 2013, 3.

9. Kitaoka Shinichi, *Gurobaru pureiya toshite no Nihon* (Tokyo: NTT shuppan, 2010).

10. *Asahi Shimbun*, December 11, 2013, 1.

11. *Asahi Shimbun*, December 8, 2013, 2.

12. *Sankei Shimbun*, December 8, 2013, 1.

13. *Sankei Shimbun*, December 11, 2013, 1.

14. *Sankei Shimbun*, December 11, 2013, 3.

15. Suzuki Nobuhiro, "TPP de idenshi kumikae shokuhin ga tairyoni ryunyu-suru," in *Nihon no ronten 2014* (Tokyo: Bungei shunju, 2013), 42–43.

16. *Yomiuri Shimbun*, December 11, 2013, 3.

17. *Yomiuri Shimbun*, November 30, 2013, 9.

18. *Yomiuri Shimbun*, December 10, 2013, 1.

19. *Asahi Shimbun*, December 11, 2013, 8.

20. Abe Shinzo, *Atarashii kuni e, utsukushii kuni e kanzenban* (Tokyo: Bunshun shinsho, 2013).

21. Murotani Katsumi, *Bokanron* (Tokyo: Sankei shimbun shuppan, 2013).

CHAPTER 19

Review Article,* October 2014: Civilizational Polarization and Japan

Gilbert Rozman

What do the battle for Ukraine, the demonstrations in Hong Kong, and the barrage of attacks on *Asahi Shimbun* have in common? In the fall of 2014 all of them are serving to polarize great powers around civilizational themes and to lessen the prospect for mutual understanding in the Asia-Pacific region. The first two of these issues have blanketed the news and confirm widespread expectations that Russia and China are increasingly under the sway of the legacy of traditional communism, but the turmoil under way in Japan is much less understood. That is the subject of the review article.

Obsessed with fundamental civilizational contrasts rather than specific interests, polarization is occurring within nations as well as between nations. This is true in the United States with its increasingly unbridgeable gap between Republicans and Democrats, in Russia fueled by demonization of Ukraine and any who would agree with it or with Russia joining a "common European home," and in China aroused by parallel demonization of the public in Hong Kong and any who would agree with it or with China allowing a civil society to pursue democratic rights. There is another case of polarization, which likewise has powerful ramifications for the international relations of the Asia-Pacific region. It is Japan in 2013–2014, and, fortunately, a book by Tokuyama Yoshio, which was published in August, documents how that is occurring with divisions over national security placed squarely in the forefront. At the end of September, *Yomiuri Shimbun* published its own book on the *Asahi Shimbun* coverage of the "comfort women" issue, further fueling the uproar over this divisive matter. In the fall of 2014 this single issue has come to symbolize Abe's polarization of Japan.

Why should *Asahi Shimbun*'s acknowledgment of a mistake, which other media also made, in accepting one man's testimony about the "comfort women" galvanize the Japanese public over the past few months with hundreds of articles on it, comments by national leaders shaping the discourse, and the *"Asahi* question" deserving to be included on a list with Ukraine and Hong Kong as tests of national identity and ties to the outside world? The reason is that the attacks on this newspaper have become the frontline in the battle over defining "normal Japan," bringing the postwar order to an end, and deciding how to deal with South Korea—the bell weather country in Japan's "reentry into Asia." As Tokuyama makes clear, the battle over reporting has been intensifying since Abe took office nearly two years ago. It has come to a head in the second half of 2014 around an issue of seemingly minor historical significance, even as this is widely recognized to have great relevance for foreign policy, political prospects, the future of Japanese media, and the reconstruction of national identity.

Under Abe Shinzo Japan has become much more deeply polarized, argues Tokuyama with extensive documentation from the six major newspapers of Japan on issues linked to national security. His argument opens our eyes to a conclusion of broader impact: since Obama, Putin, and Xi have gained their current posts, harsh attacks on opponents as civilizational traitors (in the case of the United States, unlike the other cases, they have been launched by the opposition to national leadership) have been accompanied by sharp accusations against reputed civilizational antagonists abroad. For Putin Ukraine serves as an excuse to crush civil society—the civilizational threat at home. For Xi Hong Kong, at least, gives him an added pretext to crush civil society. The case of Japan is particularly clear-cut because of the detailed media articles and editorials that Tokuyama cites, revealing the struggle of the opposition in a manner inconceivable in the authoritarian environments of China and Russia and with less clarity about how the struggle will end, given the countervailing media and opinion.

The core theme in narratives on national identity is civilization. This was evident in Japan in the 1970s–1980s upsurge of discussions about *Nihonjinron* (Japaneseness or Japanese civilization), in Russia in the simultaneous fascination (that challenged but, in a way, also reinforced Soviet identity) with earlier *Russkaia sivilizatsiia* (Russian civilization) followed in the 1990s with the rising obsession with the *Russkaia ideia* (Russian idea), and in China in the 2000s–2010s in the popularity of *wenminglun* (theory of civilization), which gave rise to a focus on the "China Dream" (*Zhongguo meng*). This review article does not compare these trends. Instead, it concentrates on divisions now reaching a boil about recovering the spirit of Japanese civilization. In each, it is assumed that domestic and foreign enemies are blocking the recovery.

The Sequence of Debate in Japanese Newspapers in 2013–2014

For six decades Japan's right wing, whose most well-known leader in the 1950s was Kishi Nobusuke, grandfather of Abe, has sought to revise the Constitution, seen as an alien imposition by the US occupation. It is understood as turning Japan into a state in direct opposition to its own past, a "peace state" with no ability to defend its own sovereignty, a state with a masochistic view of its own national identity, and a great power unable to pursue a normal foreign policy due to "bashing" by some countries and "passing" by others. Naturally, the opening theme from the late 2012 launch of the Abe administration was revision of the Constitution. According to Tokuyama, this struggle played out prior to the July 21, 2013, Upper House elections, focusing on Article 96, which stipulates the process for amendments. *Yomiuri Shimbun* and *Sankei Shimbun* clashed with *Asahi Shimbun*, *Mainichi Shimbun*, and *Tokyo Shimbun* (each side having a combined circulation of 11.5 million), as public opinion proved reluctant to support a shift from a two-thirds majority to a simple majority. Constitutional law experts weighed in with warnings of the weakening of legislative power that would occur, Abe was forced to change his tone, recognizing he had lost this fight.

The second battle detailed by Tokuyama was the fight over a secrecy law, which passed the Upper House on December 6. Seen as dealing with defense, foreign policy, intelligence, and terrorism, the proposed law aroused a heated debate. On one side, there was alarm that it would allow the state to control information at will, undermining the constitutional principle of the people's sovereignty. On the other, it was deemed to be essential for national security, assuring allies and partners that Japan would not leak sensitive information. Much of the debate centered on steps to counter China's aggressive maritime policies as the United States was weakening as a check on China.

Although the *Nikkei Shimbun* and local newspapers largely joined in opposing the law, leaving only *Yomiuri* and *Sankei* in favor, LDP success in getting its coalition partner Komeito to agree led to the passage of the law. Charging that this was a threat to the Constitution and the three-way division of power, *Asahi* took the lead in warning that this was a blow to Japan's democracy. Fear of the abuse of power and sharp limitations on freedom of information left a bitter aftertaste even as the United States was seen as welcoming a vital step in strengthening the alliance.

Coverage in the first half of 2014 was polarized over the issue of Japan's right of collective self-defense. With the establishment of the National Security Council and the issuance of a National Security Strategy in December 2013, *Yomiuri* and *Sankei* were delighted with these historic changes in postwar Japan's defense policy. In contrast, *Mainichi*, *Asahi*, and *Tokyo Shimbun* warned of a loss of balance and policy tilting unduly to the military, making dialogue more difficult. On this issue *Nikkei* took Abe's side. Making an exclusive case in *Yomiuri* for Abe's Advisory Panel on the Reconstruction of the Legal Basis for Security was Kitaoka Shinichi, who argued that recognizing the legal right to collective self-defense would have important influence on Japan's security, for example, facilitating maritime transport of energy from the Middle East. Complicating the debate, however, was the reaction to Abe's December 26 visit to the Yasukuni Shrine. Again, *Asahi* led the way, linking the two issues—historical revisionism and constitutional revisionism over issues of war and peace. *Sankei* was insistent on separating the two, asserting that Abe's move on defense was defining his "pro-active contribution to peace" to give Japan a leading role in international society. The focus turned to Komeito to see how Abe's plans would be modified, as media polarization continued. While *Nikkei* was positive about the change, its tone was much more restrained than *Sankei's*. A cabinet statement on July 1 moved this forward, but opponents were alarmed that constitutionalism was put in jeopardy by a change of interpretation without Japan following any amendment procedure. One important reason for pronounced opposition on the left and in the majority of the public is what is seen as the illegitimate method to which Abe resorted—a cabinet decision to change the interpretation of Article 9. Given the deep distrust in other changes Abe is perceived as championing, his ability to proceed with observance of constitutional procedures has raised red flags, which contribute to polarization.

In the background in 2014 were media differences over the Yasukuni Shrine visits, the role of NHK under a new politicized director, and nuclear power. It seemed as if Japan was isolated, as the United States through the winter had joined China and South Korea in criticism, as Russia and the EU questioned Abe's visit to the Yasukuni Shrine too. Only *Sankei Shimbun* supported the timing of this visit. Yet, before long the issue lost force in Japan. One reason was Abe's agreement under pressure from Obama to pledge not to revoke the Kono statement, making possible a three-way summit with Park Geun-hye in late March and greatly reducing concern about a split between allies. Indeed, Obama's late April visit to Tokyo and other countries on the front line with China proved to be a big relief, as he indicated that Article 5 of the bilateral security treaty applies to the territorial dispute with China, supported the right of collective self-defense, made clear that a new type of major power relations with China would not mean sacrificing allies, and demonstrated that there is three-way unity on complete denuclearization of North Korea. Another was the inclination of critics to equate militarism with historical revisionism, a step too far for objective viewers. A third explanation is the backlash against the worldwide campaign by China in the first months of the year to demonize Japan, equating it with Nazi Germany and Abe's actions with a return to pre-1945 Japan. Japanese newspapers split in their coverage of China with *Asahi Shimbun* warning that Abe's security moves were raising tensions with China and increasing the chances of war, while supporters of Abe insisted that his moves were the way to reduce those chances.

Forging trust became the focus, but those who insisted that Washington is the necessary target split with those who did not want to burn bridges with Beijing. On the left, fear was growing that a single cabinet could lead Japan into war and Japan was on a slippery slope leading even to conscription. They worried about the fate of democracy, pacifism, and constitutionalism—the pillars of postwar identity. The combination of moves by Abe had left a widespread feeling of alienation. With *Nikkei* joining *Yomiuri* and *Sankei* on security and nuclear energy, the other side did not have a common understanding of the national identity they were seeking, even as the momentum had largely swung to their side. In the meantime, debates over TPP found the national newspapers largely in agreement that it should go forward, without specifics on what compromises should be made, while local papers raised more objections, often representing rural interests. Suddenly in August attention shifted to the *Asahi* question, adding a rougher edge to the ongoing polarization.

The *Asahi* Question and the "Comfort Women" Issue

The *Yomiuri Shimbun* book picks up where Tokuyama stopped, carrying the story forward to August and September 2014 but presenting only one side of the debate.

Making one newspaper's coverage of one issue the focus, it connects in one overall narrative historical misunderstanding, media disinformation, political cowardice, the denigration of Japan before the world damaging its international relations, and the lack of pride in Japanese national identity. The essence of the argument is that the repeated "scoops" by *Asahi Shimbun* in the 1980s and 1990s on the "comfort women" issue—to the point that Asahi on its 115th anniversary in 1994 boasted of its success in influencing Prime Minister Miyazawa's visit to South Korea and official apology and the UN Human Rights Commission taking up the issue—dealt a devastating blow to Japan at a critical juncture at the dawn of a new era with South Korea. Its "comfort women" reporting, peaking in the first half of the 1990s, is seen as having had a decisive influence, arousing the public in South Korea to demand repeated apologies and compensation, provoking a US House resolution and parliamentary resolution in Europe, serving China's strategy of containing Japan, damaging the national interests of Japan, and slandering the good name of the Japanese people.

The *Yomiuri* editorial board, as others in Japan who have wielded the sledgehammer of this issue against *Asahi*, fails to weigh the impact of one piece of evidence that *Asahi* discovered to have been false leading to its early August apology, against all of the other evidence for the existence of sex slaves, their number, and how they were recruited. It distorts the impact of the efforts by Japanese officials with support from the media to find common ground with South Korea, leading to the historic Obuchi-Kim Dae-jung summit in 1998, as if they were the cause of South Korean anger, not an effort to assuage it. At the root of the problem is how critics of *Asahi* treat Japan's approach to the period ending in 1945, as if Japan should be proud of its behavior at that time, should treat Japanese who criticize it as hostile to Japanese civilization, and should even expect to be welcomed into the international community and Asia on the basis of repudiation of the Kono statement and, presumably, other official statements that had led the world to view Japan in a particular light. Rather than be conscious of how much of the world was reassured about Japan's postwar and post–Cold War direction from such statements, the book blames *Asahi* for putting Japan in a bad light. Claiming to restore a little of what *Asahi* had destroyed, the book totally misses the actual impact of this kind of attack on the outside world. Not only does its intemperance increase polarization inside Japan, demonizing the other side, it also deepens the divide with South Korea, which as recently as 2011 was negotiating a compromise resolution of the "comfort women" issue, plays into China's arguments about Japan, and increases the potential that historical memory will become a big problem for Japan's relations with the West. Revisionism trumps internationalism.

Polarization has deepened in Japan in 2013–2014. The left failed to grasp the importance of national security, doing its part to widen the divide. Lately, the right has seized the offensive, failing to recognize the wider impact of historical memory, intent on making this a divide between good and evil. Japanese civilization is at stake with the left appealing to the legacy of the postwar era, and the right now consciously invoking the wartime era. Neither side has helped Japan's cause in Asia or in the international community. The right is hoping that shared national security objectives will lead allies and partners to overlook its revisionist obsession. The left is appealing for shared historical memories to lead other countries to oppose Abe's security agenda. In this reasoning, the United States and China loom large, but it is South Korea that is serving as the prime testing grounds to no good effect so far.

Note

* Tokuyama Yoshio, *Abe kantei to Shimbun "nikyokka suru hodo" no kiki* (Tokyo: Shueisha shinsho, 2014); Yomiuri Shimbun henshukyoku, *Asahi "ianfu" hodo* (Tokyo: Chuko shincho La Clef, 2014).

CHAPTER 20

Realism versus Revisionism in Abe's Foreign Policy in 2014

Gilbert Rozman

In 2013 Abe Shinzo was regarded as putting realism ahead of revisionism until he visited the Yasukuni Shrine at year's end. In 2014 despite far-reaching support of realism, the balance shifted in favor of revisionism, I conclude, based especially on articles in the Japanese press on foreign policy and on sensitive themes that bear heavily on what other countries think about Japan. In the struggle against China, realism takes center stage, combining a stronger alliance with the United States, new security triangularity with Australia or India as well as maritime countries in Southeast Asia, and minimization of historical memory in favor of joint declarations of universal values. However, in the struggle with South Korea, which took priority in 2014, Japanese conservatives prioritized revisionist rejuvenation at the expense of realism. Their obsession with the "comfort women" issue vilified *Asahi Shimbun* and the progressive community at home, demonized South Korea as an emotional nation incapable of shaking off its "hate" for Japan, and targeted the United States as ready to be persuaded that for the sake of realism it should not be concerned about how Japan revises its worldview on history. The debate in Japan is being driven by what used to be considered the "far right" and is now the conservative bloc in both politics and the media. Given the seventieth anniversary of the end of World War II and the fiftieth anniversary of normalization of diplomatic relations with South Korea in 2015, the scene is set for a "history war" fought primarily on the battlefield of US opinion with, for example, *The New York Times* seen as an enemy and congressional conservatives and the security community considered to be amenable following a public relations blitz.

The meaning of realism and revisionism in Japan is readily discernible in the 2010s. Realism means building a stronger alliance with the United States, giving priority to strategic relations with front-line states facing North Korea and China, and creating a strong deterrent capacity guided by a National Security Council, which asserts the right to collective self-defense. Revisionism means reinterpreting the history of World War II and Japan's expansionist role in Asia, approving symbols such as visits to the Yasukuni Shrine, denying the symbols of Japanese inhumane behavior, and reinterpreting the Tokyo Tribunal and negative judgments of past conduct. In theory, no trade off is needed between greater realism and greater revisionism. Yet, in light of the sensitivity of Japan's historical memories to South Koreans, Chinese, some in Southeast Asia, and even Americans, the more the revisionist agenda is in the forefront, the harder it is to get cooperation on a realist agenda. Thus, moves by Abe and those seen as in his camp are scrutinized for how they weigh this balance.

The growing presence in Abe's successive cabinets of vocal representatives of right-wing causes amplifies the voice of the publications of similar persuasion, such as *Sankei Shimbun* and *Yomiuri Shimbun* (as the latter's orientation on issues such as South Korea has shifted to the extreme). In turn, Japan's left-wing representatives and much of public opinion have muddied the differences between a centrist, realist approach to an increasingly dangerous region and a revisionist approach to history that serves to isolate Japan and can add to that danger. As a result, the far-reaching struggle between realism and revisionism was misconstrued on both sides, as if the two go hand in hand and are not in conflict. Here the contradiction between these two approaches is showcased with an emphasis on perceptions of South Korea. No country more than it exemplifies the struggle between these clashing orientations.

What determines the balance between realism and revisionism in Japanese policies? The predilections of the prime minister—in this case Abe—are naturally important. Every indication is that he wants it both ways, conducting policies that have long been the priority of realists and also obsessing about historical memory, patriotic upbringing, and other themes of great significance to revisionists. The cabinet and the impact of coalition partners—that is, Komeito—matter. After all, Japan's top leader lacks the personal clout of the US president or of the other leaders in Northeast Asia.

Another factor is public opinion, critical in a democracy where the nation's pulse is constantly being taken. Finally, there is the factor of the internal environment, that is, the economy determines the availability of resources, and the external environment, the degree of dependency on one country being one concern. Abe's choices reflect a mixture of these variables in 2013–2014 and the prospects for them in the year 2015.

Conservative versus Progressive Views in 2014

In 2014 Abe showcased his successes in Japan's long overdue shift to a proactive, security-centered approach; the US alliance was seen to be strengthened through improved prospects for collective defense, intelligence sharing, and multilateral alliances and partnerships; and Abe found it possible to hold summits—however abnormal their appearance—with President Park Geun-hye in March and with President Xi Jinping in November. TPP talks were proceeding, despite moments of doubt, in pursuit of strategic as well as economic objectives. Conservative media trumpeted that Japan is back from two decades of weak leadership, 70 years of abnormal passivity, masochistic denial of self-pride, and indifference to national security. The argument permeating the media is that Japan's wartime behavior—its honor—has been smeared without evidence by anti-Japanese groups, including those led by Chinese-Americans and Korean-Americans. The retraction by *Asahi Shimbun* is treated not as proof that one source was exposed as invalid or even that one issue was reported inaccurately, but that the entire narrative about the war propagated in China, South Korea, and the West is erroneous. The Nanjing massacre, the medical atrocities of Unit 731, and all other charges against Japan's conduct at the time are seen as falling like dominos with the retracted evidence in August 2014.

Japanese progressives generally fail to make a distinction between revisionism and realism too. An editorial in *Asahi Shimbun* argued that in light of China's military expansionism and maritime advance, it supports a stronger Japan-US alliance, but that this is not directly connected to Japan's pursuit of the right of collective self-defense or to amending Article 9 of the Constitution, which would change the nature of Japan's pacifism that puts a distance between Japan and the disputes occurring abroad.[1] Whether collective self-defense that breaks with pacifist avoidance of any role in foreign conflicts or historical revisionism, this newspaper is firmly opposed. It also warns that given China's strong reaction to the fact that after the brief Abe-Xi summit Japan has not changed its posture on historical consciousness and the island dispute, Japan should be prepared for its historical consciousness to be the focus of the seventieth anniversary commemorations in China. Distrust in Abe is not diminishing.[2]

The difficulty for Japanese progressives is that they have defined themselves as an intelligentsia—cultural and academic—by reliance on the media they read, notably *Asahi Shimbun*, and by opposition to revisionism and realism, as if they are coupled. Japan is no longer split into highbrow and mass culture, the collapse of the unions as a bulwark of the progressive cause has altered the political scene, and the *"Asahi* brand" is dying, Takeuchi Yo writes, but embracing realism is hard. It is not made any easier by Abe's insistence on revisionism along with it. Through the 2000s, even in the troubled middle years, talk of an East Asian community (*Higashiajia kyodotai*) allowed stress on future shared values to mitigate tensions over historical concerns. Various "pipes" such as former ambassadors and progressive media showcased this theme,[3] but officials on all sides and conservatives at least paid lip service to it. In 2014 there was no prospect of shared values with China and no sign that forces on the right saw any reason to tone down revisionism to achieve foreign policy goals. Media on the right felt unrestrained, while those on the left had scant basis for hope.

The problem has not only been the traditional bifurcation of the Japanese media with scant evidence of the middle ground. Under Abe, aspects of national identity that complicate trust in the United States, support for the principles of international society, and appreciation of the complexities of the reordering of Asia have become more pronounced. In the successive November summits in Beijing, Naypyidaw, and Brisbane, Japanese had an opportunity to take stock of the many currents at work in the Asia-Pacific region. They did so with undue ambivalence about Obama's policies, setting the tone for unrealistic appraisals of what Japan could expect to accomplish.

The Shadow of the United States

A review of what the United States, arguably, has done for Japan's foreign policies in 2014 requires a lengthy list. It reduced some of the spillover in the first months of the year from Abe's Yasukuni visit by expressing disappointment, demonstrating that revisionism occurs at the expense of realism, not as its natural accompaniment. Obama's team worked tirelessly to narrow the divide between Abe and Park, even to the point of holding their hands at a three-way summit in The Hague when they had no intention of compromising with each other to make a bilateral summit possible. In Japan's debates on the secrecy law and the right of collective self-defense, appeals to their importance for the alliance were critical, given the high popularity and wide acceptance of the alliance as vital to Japan's security. Even if Abe's motives and some of the proposed contents of these policy changes raised suspicions among a sizable number of Japanese, the realist case for supporting an indispensable ally served to move these initiatives forward. Japan managed to join in sanctions against Russia over its conduct in Ukraine, despite disclaimers that this was done only under the pressure of the United States and the need to cooperate in the G-7, again avoiding an image of breaking with the realism of the international community. Finally, the fact that Xi Jinping met Abe after months of uncertainty, given the conditions Xi had set for a meeting to go forward on the sidelines of APEC, was in no small measure due to the efforts of US officials working with China to prevent the fallout from the two not meeting and to establish a China-Japan mechanism to prevent escalation from an unforeseen incident at sea. We can add to this list Abe's success in "piggybacking" on US overtures to India, Australia, and countries in Southeast Asia, "rebalancing" to the South China Sea and the Indian Ocean on the heels of Obama's many initiatives.

Reading Japanese newspapers, however, one would be hard pressed to find credit to the Obama administration for any of these accomplishments pursued with Abe. On the contrary, the prevailing images are of: Japan "passing" placing more importance on relations with China; "gaiatsu" or pressuring Abe to do things not in the national interest of Japan; "disrespect" in failing to have high officials and the president give suitable credit to America's "indispensable ally" in the Asia-Pacific region; and lack of professionalism and consistency in strategizing over developments in the region.

Criticisms of the United States focused on the triangle with China. In *Tokyo Shimbun* Kimura Taro called for sweeping away Obama's natural security team, starting with Susan Rice, following the Republican victories. He detailed a string of failures for US foreign policy since she assumed her post, which left allies, including Japan, uneasy, especially after the US expression of disappointment following Abe's Yasukuni visit and the wavering in the US position on the Senkaku islands and TPP. Japan, he said, has become uneasy about relying on the United States since Rice arrived and gave priority to China,[4] as in her September trip to Beijing, where she accepted Xi's idea of two superpowers, setting back relations with Japan. Kimura wondered if

Obama's electoral setback might lead to a policy change, that is, to one backed by Republicans.

On November 24 *Gendai Business* portrayed the essence of the back-to-back APEC summit and state visit of Obama to China as part of China's effort to establish a "new type of major power relations." Inviting Obama to dine in *Zhongnanhai*, as Mao did for Nixon in 1972, broke with recent precedent and reflected the goal of dividing the Pacific Ocean into two, leaving East Asia to China. Preoccupied with the Ukrainian crisis, the rise of the Islamic State, and Iran's nuclear program, Obama must rely on China to prevent friction in its neighborhood was the message. In return, trade would rise beyond its current USD 520 billion while cooperation would advance on global warming, Ebola, terror, the Iranian and North Korean nuclear programs, and Syria. Both countries' national interests would be served, Asia-Pacific peace and security would be maintained, and prosperity would be enhanced, Xi was asserting.

Articles noting the US decline appear often. One *Yomiuri* piece noted that countries have lost their fear of Obama and no longer expect US hegemony, and even pro-US states, one-by-one, are distancing themselves from the United States. In describing a relative vacuum, the paper discusses a need to strengthen Japan and the importance of restoring its honor, sullied by *Asahi Shimbun*, by reexamining history.[5] Elsewhere one reads that the world is losing its policeman, order has broken down across the Middle East, and the situation in East Asia is perilous, which requires greater US reliance on Japan. The turnaround in the US economy and confidence was slow to register, even in late 2014, as Japan's economy slumped. Emboldening Japan as a force for reshaping East Asia took priority over joining in a US-led regional order.

The more the US government recognizes the threat from China, the more Japanese rightists perceive an opening for revisionism since Washington needs Tokyo more, beginning with its shift to collective self-defense. One example is a *Sankei Shimbun* article on the congressional executive commission on China report released the day before, which was seen as reducing pressure on Japan.[6] The main battlefront in the "history war" is the United States, *Sankei* noted,[7] stressing a struggle in Congress, where Korean Americans are seen as making inroads in an "anti-Japan" conspiracy.

Having given Lee Myung-bak and Park Geun-hye the honor of speaking to a joint session of Congress and with Republicans now in charge, their leaders are likely to invite Abe as well when he visits the United

States, perhaps in the spring of 2015. It is likely to be a time of celebration of TPP passage or agreement on joint guidelines. Rallying behind Japan is an obvious response to China's efforts to isolate the state and split the US-Japanese alliance. No doubt, Abe would use the occasion to shape the debate over the seventieth anniversary. He is likely to be walking a tightrope, seeking to earn praise as a realist while doing little to diminish his quest for revisionism.

Yomiuri Shimbun at year's end noted concern in Europe and the United States over Abe's historical consciousness but sought a more strategic response. As the US-led international order faces the seventieth anniversary of the war's end, it added, care must be taken to ensure balance in the international arena, where Japan is expressing its concern over the dangerous moves of Xi Jinping, its forces are contributing through collective self-defense and other moves, and it is supporting the Philippines and Vietnam in a joint international struggle.[8] Forget revisionism, accept Japan's realism.

Asahi Shimbun claims to support realism in the form of a stronger Japan-US alliance and warns against revisionism, as a threat to US trust in Japan and to relations with South Korea. Yet, its warnings against visits to the Yasukuni Shrine or denials of any responsibility for wartime excesses are coupled with warnings against one-sidedly leaning on military power, as if Japan does not need to affirm the right of collective self-defense, to expand its defense budget, and to start exporting arms due to the new regional and international security environment.[9] This newspaper frequently carries views of US, South Korean, and Chinese specialists, which cast doubt on the views of Japanese conservatives. For example, on December 8 it reported on Li Wei as well as Sheila Smith, the former as head of CASS's Japan Institute arguing that Abe has disappointed China with his revisionist moves, including his visit to Yasukuni and his vagueness on what is meant by aggression—concerns that are shared with US scholars—and it is worried about what Abe will do to mark the seventieth anniversary in contrast to the Murayama statement. Li interpreted the Xi-Abe summit as just an opportunity to express each side's positions, not as in past bilateral summits a step toward improving relations.[10] Progressives were looking for a silver lining behind the clouds hovering over East Asian security ties in order to expose revisionism as the real culprit and to discredit the need for realism, which would change the character of Japan. They tried to find signs of a softening in China's posture, but this often proved difficult, even as they cited Chinese views blaming Abe's thinking.

South Korea's Place in Japan's Pursuit of Revisionism

The argument of Japanese conservatives is that Japan can have revisionism as well as realism without having to make substantial tradeoffs. This is premised on views of three countries, above all. The assumption about the United States is that those who count (the Congress and the security community more than Obama and some Democrats) care much more about realism and, when pushed, will let revisionism go ahead with little more than a murmur. China, in turn, is seen as so hostile to the realism of Japan that its animosity toward revisionism does not change its reaction very much. It follows that the United States will focus on a long-sought, tightening alliance, and China will demonize Japan as an unredeemed enemy. That leaves the interpretation of South Korea as the principal test of whether there is a cost to the strategy of pursuing realism and revisionism simultaneously. Five arguments are made by those who insist that South Korea does not pose a serious cost on Japan: (1) it is hopelessly anti-Japan, so that both realist and revisionist moves matter little; (2) it is bound to the United States, so that realism and US pressure will win the day; (3) it is shifting to China and soft on North Korea, so that Japan's realism is rejected: (4) it is troubled economically and socially, so that it is of little consequence for Japan; and (5) South Korea is in danger of isolation, so that it must yield to Japan's demands.

South Korea is an economic competitor, soon to be connected to China through the strengthened bonds of an FTA. In electronics and automobiles, competition is most intense. The devaluation of the yen has improved Japan's position, but its identity would be shaken if South Korea were to overtake it in some visible manner, as China recently did in GDP. Articles that put South Korea's economy in a doubtful light keep this challenge under control. On November 21 *Yomiuri Shimbun* explained that Park Geun-hye is having little success with her goal of "economic democratization"—one shared by recent presidents to little avail, as the country's dependence on *chaebol* keeps growing—in the face of public concern over widening inequality. Superior college students heavily focus on income security through chaebol and public jobs. The article concludes with the argument that the fundamental problem is the loss of entrepreneurial spirit in South Korean society. Such negativity about the South has the effect of avoiding recognition of the need for realism toward it.[11]

Kimura Kan has carefully examined the impact of *Asahi Shimbun*'s coverage in the 1990s of the "comfort women" issue on South Korea, focusing on ten newspapers in the South. Contrary to the assertions by Japanese conservatives, he found only a limited impact in South Korea, although he sees it having a big impact inside Japan.[12] In contrast, Chief Cabinet Secretary Suga insisted that the *Asahi* reporting on the "comfort women" has had extraordinary influence.[13] He claimed that in the past the Japanese diplomat establishment had been relatively silent against arguments that the women had been coerced, but now there is a big change as ambassadors and those below them will respond with all their energy as public relations budgeting has been doubled, and Japan's position will be clearly presented in the United States.

Choosing to prioritize the "comfort women" issue in 2014—even more so after the campaign against *Asahi Shimbun* gathered steam in August—Japanese conservatives made South Korea (not China) their main target and the United States the principal battleground, as they fought against South Korea and persons of Korean descent, who were criticized for anti-Japan lobbying in resisting Abe's revisionism.[14]

Just as the August demonization of *Asahi Shimbun* was putting the "comfort women" issue on center stage, setting back relations, the arrest of a *Sankei Shimbun* reporter for libel in Seoul gave Japanese a strong reason to challenge South Korea's values. As well-known writer Sato Masaru wrote, for a country with shared values of freedom, democracy, and a market economy, how could it behave in such a manner?[15] Others note that postwar Japan has defended such values and been a model peaceful state; so how can South Korea as well as China disregard these factors and criticize visits to Yasukuni Shrine?[16] Discrediting South Korea's realism helps the case against its opposition to revisionism, arguing that it along with China is the exception to the overwhelming support around the Pacific Ocean for Japan's right of collective self-defense. This is attributed to an irrational emotionalism toward Japan as well as a lack of realism in recognizing the challenge from China.[17] South Korea is an outlier.

The attacks against *Asahi Shimbun* were construed as the removal of an albatross around Japan's neck both domestically and in relations with South Korea above all. Now Japan could look ahead to true normalization focused on the future, it was said, as a "normal Japan" no longer would consider criticisms over history. Its time on the defense was over, a *Yomiuri Shimbun* editorial implied.[18] Maintaining the claim that Japan is a "peace country" (*heiwa kokka*), Abe at the United Nations in September insisted that this identity persists without pointing to continuation of apologies.[19]

The Resurgence of History

It is hard to look past 1945 in East Asia when Japanese conservatives keep pointing to the need to reassess views of it, South Koreans keep looking at Japan in terms of it, and Chinese dramas and even academic narratives keep recalling it. Abe and Xi oddly are conspiring to keep the spotlight on the region's past—each insisting that this is the path to national rejuvenation as well as realist policies—while Park is not adverse to prioritizing the past over the future in regard to relations with Japan.

In a recent book published by *Sankei Shimbun*, *The History War* (*rekishisen*), it is argued that the main enemy is China, the battleground is America, China's partner is South Korea, the aim is to isolate Japan, the key issue is "comfort women," a major target is Congress, leftists in Japan long strove to put their country under siege, Japan's Foreign Ministry and government failed to counterattack, Japan is in danger, and only concerted efforts by its government and people will allow it to win this necessary war.[20] All of this is explained as a result of the influential falsifications about the "comfort women" disseminated around the world by *Asahi Shimbun*.

At the September 3, 2014 ceremonies (hereafter to be a three-day commemoration) on the sixty-ninth anniversary of the war's end with all seven members of the leadership in attendance, Xi Jinping made it clear that, even as he meets with Japanese officials and talks of improving relations, he is preparing to increase pressure over history for the seventieth anniversary.[21] As a victor in the war, China claims now to be defending the postwar international order, which Japan supposedly is threatening.[22] Concern is mounting in Japan that China will use this anniversary against their state.

Sankei Shimbun warned that Chinese Americans are organizing to make the seventieth anniversary all around America a time for "anti-Japanese" activities, including on the stage of the US Congress.[23] The battle in the United States is heating up, readers read. Abe wants summits with Park and Xi (Park seemed to okay a three-way summit early in 2015 after foreign ministers would meet, but Xi's response was doubtful), a triangle with India involving the United States or Australia (but Modi was cautious when they met in Brisbane),[24] and breakthroughs with Putin and Kim Jung-un. He is also interested in a triumphant visit to Washington in 2015, showcasing the US Congress' enthusiastic acceptance of Japan on the occasion of the seventieth anniversary with clear support for its realist shift away from the "postwar regime" and sufficient agreement for his wording of a statement on history to permit his revisionist agenda to go forward. For Japan's right wing the Abe-Xi summit is one more step to make Park and the media in South Korea more realistic about relations with Japan.[25] Japan can have its shift to revisionism, but others must be more realist in accepting it. The right focuses on South Korea's isolation, not Japan's, and its need to back down.[26]

Conclusion

Abe's election victory in December 2014 did not free him of the constraints on both his realist and revisionist agendas. Komeito is stronger as a coalition partner. Public opinion remains wary of both of his agendas. The internal economic environment is gloomy. The external environment is more favorable to the United States, less so for an initiative to Russia, and more confining due to improved Sino-US relations. While November US congressional elections boosted Republicans, prospects for success in a Japanese public relations blitz in the United States have not risen significantly. To the limited degree that Sino-Japanese and South Korean-Japanese relations slightly improved over the past few months, pressures for a realist approach are growing.

Japan's weakness is rarely recognized. US decline is exaggerated. Japan's diplomacy in Asia is presented with hyperbolic hopes. Japan's salience for the international order and global balance is overstated. Based on these extreme assumptions, the case is made that South Korea will have to yield to Japan for realist reasons and Japan should not make concessions on "comfort women" or Takeshima.[27] Critical to this line of reasoning is the assumption that the United States is so dependent on Japan, so willing to pressure South Korea, and so focused on a "China threat" that its leaders (perhaps not Obama, who is not well trusted) will have to side with Japan. In support of this is a persistent effort to fudge the difference between revisionism and realism, as both emphasize strengthening the Japan-US alliance and widening its reach through triangular relations with India and Australia, "isolating" China. What is distinctive about revisionist thinking is the stress on isolating South Korea too.[28] At the same time, some charge that Obama is naive or that the United States needs to avail itself of the seventieth anniversary to stop thinking of itself as the winner and Japan as the loser, and to change its consciousness (particularly in light of recent articles in *The New York Times* critical of Japan's revisionism) regarding Japan.[29]

Okura Kazuo, former ambassador to South Korea, offers four reasons why the gulf in Japan-South Korean relations has widened so far: (1) the media on both sides; (2) the divide in historical consciousness; (3) the difference in social structure; and (4) the change in the power balance in East Asia, which is the background that makes it extremely difficult to resolve politically and strategically the "comfort women" issue. He points to *Donga Ilbo*'s coverage of a planned ceremony in the Lotte Hotel in July by the Japanese Embassy in Seoul for the establishment of the Self-Defense Forces, which led to a flood of opposing phone calls and the hotel deciding to cancel what had been an annual event. Other Korean media as well make it hard to compromise with the United States or Japan, in a tradition of extremely distorted coverage as if Japan is trying to defend its militarist era. TV announcers too are opinionated and lead public opinion, argues Okura. Japanese media too, led by *Asahi*, oppose the conservative regime and foster a mentality of traditional, emotional opposition. Yet, Okura holds that *Asahi's* misreporting did not necessarily harm Japan's interests and he sees bias in the fact that many other media are repeatedly, hysterically bashing it.

His second point is that a divisive mentality is interfering with greater cooperation. On the third point, Okura recalls Chun Doo-hwan's first visit by a Korean president to Japan in 1984, when he failed to support 100 percent the normalization agreement of his predecessor in 1965, opening the door to the "comfort women" issue becoming the litmus test for what was seen as real normalization, when, in fact, the wartime treatment of women should be a broad human rights concern, not a bilateral issue. Okura attributes South Korean emotional preoccupation with this issue to its social structure, where in rapid democratization citizens' movements (as well as unions) acquired great power. At the same time, as Japanese in the midst of stagnation saw China overtake their country and Korean firms Samsung and Hyundai challenging Japanese corporations a feeling spread that Japan is the victim or that South Korea already is equal and should not demand more. This leads to the fourth point that the principal reason that Japan cannot improve relations with China and South Korea is because of is the decline in its national power. Okura disagrees with the argument of many writers that Xi Jinping and Park Geun-hye are using Japan as a scapegoat to deflect attention from the weakness of their own support base. Rather, he contends that they do not see a weak Japan as a worthy strategic partner. He concludes that Japan should be careful about how it handles the "comfort women" issue, given the danger it could be seen as a strange country for its values in Europe and the United States. Instead, it should separate this issue from bilateral ties to Seoul and try to join Seoul in championing human rights for women with a fund, changing its image from that of a country that does not value the human rights of the weak and discriminates against women. Okura rejects the revisionist agenda in favor of universal values and realist recognition of declining power, which requires a more strategic approach.[30]

In late 2014 Kitaoka Shinichi made the case for realism, while both casting doubt on South Korean assertions about how many "comfort women" there were and denying that much damage was caused to Japan's image since it is well regarded by most of the world. What Kitaoka finds to be the main source of Japan bashing is its legacy of isolationist (*sakoku*) thinking, which has been used by progressives to weaken the government and leave it unable to fulfill its international responsibilities. Yet, he is careful to explain that criticism of one's own government raises state legitimacy. At a time when most of Asia is looking to Japan to play a greater security role—treating the new secrecy law and right of collective self-defense in this light—*Asahi*'s charge that Asia is concerned about the militarization of Japan is what really is disturbing, he adds, arguing that this echoes China and North Korea as well as South Korea, which is much more focused on "comfort women." Suggesting that *Asahi* should be left to correct it own past mistaken reporting, Kitaoka argues that although South Korea is now too emotional to reach an understanding with Japan ("it takes two to tango"), if Japan turns its eye to the world offering assistance to female victims (not excluding Vietnamese women who were exploited by South Koreans and Korean women exploited by Americans), then it could act in concert with South Korea.[31]

Notes

1. *Asahi Shimbun*, December 1, 2014, 8.
2. *Asahi Shimbun*, December 5, 2014, 7.
3. *Asahi Shimbun*, December 21, 2005, 15.
4. *Tokyo Shimbun*, November 9, 2104.
5. *Yomiuri Shimbun*, December 8, 2014, 24.
6. *Sankei Shimbun*, November 21, 2014, 9.
7. *Sankei Shimbun*, September 20, 2014, 1.
8. *Yomiuri Shimbun*, December 10, 2014, 15.
9. *Asahi Shimbun*, December 9, 2014.
10. *Asahi Shimbun*, December 11, 2014, 15.
11. *Yomiuri Shimbun*, November 21, 2014, 7.

12. Kimura Kan, "Asahi hodo wa jissai, Kankoku ni donoyo no eikyo o ataeta ka?" *Chuokoron*, November 2014, 74–79.

13. "Suga Yoshihige interview: Suga Kanbochokan ga kataru Abe seiken, tsugi no itte," *Chuokoron*, November 2014, 130–131.

14. *Sankei Shimbun*, August 7, 2014, 1.

15. *Sankei Shimbun*, November 7, 2014, 4.

16. *Yomiuri Shimbun*, November 6, 2014, 11.

17. *Sankei Shimbun*, August 3, 2014.

18. *Yomiuri Shimbun*, August 6, 2014, 3.

19. *Yomiuri Shimbun*, September 27, 2014, 3.

20. *Sankei Shimbun*, November 9, 2014, 10.

21. *Yomiuri Shimbun*, September 4, 2014, 7.

22. *Yomiuri Shimbun*, September 5, 2014, 3.

23. *Sankei Shimbun*, December 1, 2014, 1.

24. *Asahi Shimbun*, November 15, 2014, 4.

25. *Sankei Shimbun*, November 15, 2014; *Yomiuri Shimbun*, November 14, 2014, 1.

26. *Sankei Shimbun*, November 18, 2014, 6.

27. *Sankei Shimbun*, September 5, 2014, 7.

28. *Yomiuri Shimbun*, September 7, 2014, 1.

29. *Sankei Shimbun*, November 6, 2014, 7.

30. Okura Kazuo, "Nikkan no mizo ga koko made hirogaru honto no riyu," *Chuokoron*, November 2014, 68–73.

31. Kitaoka Shinichi, "Seifu no mukoo niwa sekai ni aru: Sakoku shiko o dassuru toki," *Chuokoron*, November 2014, 38–41.

Contributors

Dennis Blair is the chairman of the board and CEO of Sasakawa Peace foundation, United States. From January 2009 to May 2010, as director of National Intelligence, Blair led 16 national intelligence agencies. Prior to retiring from the Navy in 2002 after a career of 34 years, Admiral Blair was the Commander in Chief, US Pacific Command. He has recently written *Military Engagement: Influencing Armed Forces Worldwide to Support Democratic Transitions.*

Bong Youngshik is a senior research fellow at the Asan Institute for Policy Studies. He was an assistant professor at American University and Freeman Assistant Professor of Korean Studies at Williams College, MA;. His research focuses on the interplay between nationalism and regional security including island disputes and historical reconciliation in East Asia, anti-Americanism, the US-Korea Alliance, and North Korea's nuclear program.

Choi Kang is the vice president for research and the director of the Center for Foreign Policy and National Security at the Asan Institute for Policy Studies. He was the dean of Planning and Assessment at the Korean National Diplomatic Academy, professor and director general for American Studies at the Institute for Foreign Affairs and National Security, senior director for Policy Planning and Coordination on the National Security Council Secretariat, and a South Korean delegate to the Four-Party Talks. He writes on the ROK-US alliance, North Korean military affairs, inter-Korean relations, crisis management, and multilateral security cooperation.

Gi-Wook Shin is a professor of sociology and senior fellow at the Freeman Spogli Institute for International Studies at Stanford University, where he is director of the Walter H. Shorenstein Asia-Pacific Research Center. Shin is the author/editor of 18 books and over 50 articles, most recently *Global Talent: Skilled Labor as Social Capital in Korea.* After receiving his BA from Yonsei University, he was awarded his MA and PhD from the University of Washington.

Thomas Hubbard is a senior director and veteran leader of McLarty Associates' Asia Practice and chairman of the Korea Society in New York. A career foreign service officer for nearly 40 years, he served as US ambassador to the Republic of Korea from 2001 to 2004 and as ambassador to the Philippines from 1996 to 2000. He also served seven years in Japan and was deputy chief of mission and acting ambassador in Malaysia. He has held key Washington postings, including principal deputy assistant secretary for East Asian and Pacific affairs.

Hyun Daesong is an associate research fellow at Korea Maritime Institute. His research focuses on Korea-Japan relations and territorial disputes. He received his BA from Hankuk University of Foreign Studies and completed the doctoratal degree in international politics at Tokyo University. He was associate professor at Tokyo University and research professor at Kookmin University.

Kiichi Fujiwara is director of the Security Studies Unit, Policy Alternatives Research Center, and professor of international politics at the University of Tokyo. A graduate of the University of Tokyo, he has held positions at the University of the Philippines, the Johns Hopkins University, and the University of Bristol and was a fellow of the Woodrow Wilson International Center. His works include: *Remembering the War Is There Really a Just War? and Conditions of War.*

Kim Jiyoon is a research fellow and director of the Center for Public Opinion and Quantitative Research at the Asan Institute for Policies Studies. She was a postdoctoral research fellow at Université de Montréal. Her research interests include elections, voting behavior, and American politics. She received her BA from Yonsei University, MPP in public policy from the University of California, Berkeley, and PhD in political science from MIT.

Koichi Nakano is professor of political science at the Faculty of Liberal Arts, Sophia University with a PhD in politics from Princeton University). His research has focused on Japanese politics, including neoliberal globalization and nationalism, and the Yasukuni problem. He was the author of *Party Politics and Decentralization in Japan and France: When the Opposition Governs.*

Lee Chung Min is a professor of international relations at the Graduate School of International Studies, Yonsei University. He received his PhD in international security studies from the Fletcher School of Law and Diplomacy, Tufts University. He is currently serving as ambassador for National Security Affairs in the Korean government. He has written widely on defense and foreign policy issues in East Asia.

J. Berkshire Miller is a specialist on Northeast Asian security issues and has held a variety of positions in the private and public sector, recently as a Sasakawa Peace Foundation fellow on Japan with the Pacific Forum CSIS, where chairs a group focused on Japan-Korea relations in the context of the US "rebalance" to Asia. Miller is also a fellow on East Asia for the EastWest Institute and is working on projects related to Japan-China tensions in the East China Sea.

Nakanishi Hiroshi is a professor in the Department of Law at Kyoto University, where he also received his BA and master's degrees. He has written extensively on international politics and on Japanese foreign policy and changes to the world order as well as on the Japan-US alliance. He was a PhD candidate in the Department of History at the University of Chicago.

Park Cheol Hee, PhD at Columbia University, is a professor in the Graduate School of International Studies and director of the Institute for Japanese Studies at Seoul National University. He teaches Japanese politics, Korea-Japan relations, and international relations in East Asia. He authored *Daigishi no Tsukurarekata* (*How Japan's Dietman Is Made*) and *Jamindang Jongkwon gwa Jonhu Cheje eui Byunyong* (*LDP Politics and the Transformation of Postwar System in Japan*).

Gilbert Rozman retired from Princeton University in 2013 and started serving as editor in chief of *The Asan Forum*. He is the emeritus Musgrave professor of sociology. His writings focus on the countries of Northeast Asia, comparing countries and examining international relations. As editor, he has encouraged broad involvement by authors from the Asia-Pacific region as well as from the United States and close attention to how issues are viewed within the region.

Sheila A. Smith, an expert on Japanese politics and foreign policy, is senior fellow for Japan studies at the Council on Foreign Relations (CFR). She is the author of *Intimate Rivals: Japanese Domestic Politics and a Rising China* and Japan's New Politics and the U.S.-Japan Alliance. Prior to joining CFR in 2007, she directed a multinational research team at the East-West Center in a cross-national study of the domestic politics of the US military presence in Japan, South Korea, and the Philippines. She earned her PhD from Columbia University.

Lieutenant General **Yamaguchi Noboru** is professor at the International University of Japan. Upon graduation from the National Defense Academy of Japan he joined the GSDF as an army aviator. He received his MA from the Fletcher School and was a national security fellow at Harvard University. After serving 35 years in army uniform, he retired from active duty and joined the NDA in 2009 as a civilian professor. After the East Japan Great Earthquake, he served at the Prime Minister's Official Residence as special adviser to the Cabinet for Crisis Management.

Yuichi Hosoya is professor of international politics at Keio University, Tokyo and a member of the advisory board at Japan's National Security Council (NSC). He was a member of the Prime Minister's Advisory Panel on Reconstruction of the Legal Basis for Security (2013–2014) and a of the Prime Minister's Advisory Panel on National Security and Defense Capabilities (2013). His interests include postwar international history and Japanese foreign and security policy.

Index

CPSIA information can be obtained at www.ICGtesting.com
Printed in the USA
LVOW07*0255161115

462732LV00009B/86/P